The Internet of Elsewhere

The Internet of Elsewhere

The Emergent Effects of a Wired World

CYRUS FARIVAR

FOREWORD BY VINTON G. CERF

RUTGERS UNIVERSITY PRESS

NEW BRUNSWICK, NEW JERSEY, AND LONDON

LIBRARY OF CONGRESS CATALOGING-IN-PUBLICATION DATA

Farivar, Cyrus.

The internet of elsewhere : the emergent effects of a wired world / Cyrus Farivar ; with a foreword by Vinton G. Cerf.

p. cm

Includes bibliographical references and index.
ISBN 978–0–8135–4962–0 (hardcover : alk. paper)
  1. Internet—Social aspects—Developing countries—Case studies.
2. Internet—Political aspects—Developing countries—Case studies.
3. Information technology—Social aspects—Developing countries—Case studies.
4. Information technology—Political aspects—Developing countries—Case studies.
5. Internet—South Korea.   6. Internet—Senegal. 7. Internet—Estonia.
8. Internet—Iran.   I. Title.
  HM851.F35   2011
  303.48′3309—dc22   2010024091

A British Cataloging-in-Publication record for this book is available from the British Library.

Visit our Web site: http://rutgerspress.rutgers.edu

Manufactured in the United States of America

To my parents, Sydney and Mehrdad, who showed me the way,
and to Rebecca, who kept me going

CONTENTS

FOREWORD

VINTON G. CERF

I have just finished reading this book. It is an amazing amalgam of history, despair, triumph of the human spirit, and close-up glimpses of the complex fabric of social, political, and technological revolution. Written through a framework of countries where change has come despite all odds, it takes the reader into cultures that have absorbed, adapted, and altered the Internet fabric to fit their unique contexts.

I had the pleasure of meeting the author of this book, Cyrus Farivar, at a conference called LIFT, in 2009. I enjoyed our give and take in an interview, but until I read this book I did not fully appreciate the depth of passion that the Internet and its story in these countries had stirred in his soul.

I was reminded of another book, written in 1992 by Carl Malamud, titled *Exploring the Internet*. Malamud was already blazing historical trails in an Internet that was just making its way onto the public's radar. It was the year when I first heard people apologizing for not having an e-mail address on their business cards. Considering that networked e-mail had been invented in 1971, this was a long, twenty-year gestation! The World Wide Web, although conceived and first made concrete at CERN on Christmas Day, 1990, had not yet emerged to public consciousness. But Malamud's tales chronicled a global awakening.

Farivar has done something similar in this book, but also draws our attention to the courage it has taken in some places to pursue access to and development of Internet infrastructure in regimes where freedom of expression is the exception, not the rule. The old adage: "Where there's a will, there's a way," seems eminently appropriate to describe much of what is found in these stories.

While the Internet could not have become what it is today without the concerted efforts of millions, it is still remarkable that the success stories in this book revolve around one or perhaps a few people who dedicate themselves to planting and growing the Internet's seeds in sometimes resistant soil. These stories reinforce the notion that sometimes one person can make an enormous difference. It is a reminder that patience and persistence are often the only ways to achieve long-term goals.

In reading about the Internet in Estonia, I was impressed by the range and depth of applications developed there, especially regarding government services and electronic voting. But I was even more surprised to realize that all of this has happened in less than twenty years, since the country emerged from the Soviet Union's iron cloak in 1991. That these online applications have proven not only achievable but also sustainable gives sustenance to those with similar aspirations to change the state of their societies and to make them more open and transparent.

Economics is playing a key role in the continued spread of Internet access. Estimates of Internet use now hover just below 2 billion users, worldwide. Surprisingly, about 400 million of these users are in the People's Republic of China! There are an estimated 4.5 billion mobiles in use, perhaps 15–20 percent of which have some kind of Internet access. Device costs are coming down. Means of access include 2G, 3G, 4G, LTE, WiFi, fiber to the home, digital subscriber loops, broadband cable, WiMax, conventional forms of Ethernet, 6Lowpan, and dial-up modems, among others. College campuses and Internet cafés are often the first places that provide Internet access, although the smartphone/mobile is clearly taking the lead in countries where Internet access is still nascent.

It is hard to predict what the consequences of an increasingly networked planet will be, but one thing is certain: our ability to share information, to find information, and to use information can only be enhanced by the proliferation of Internet and related technology throughout our global society. Our descendants in the twenty-second century will read books like this one and wonder at the primitive "Estonia-age" uses of the Internet. But for us, it's as if the invention of fire was not so long ago and the wheel is yet to come.

ACKNOWLEDGMENTS

This book is dedicated to my parents, Sydney and Mehrdad Farivar, who created ample opportunities for me to explore my own elsewhere. My mother, who sadly passed away in February 2010, helped me get online for the first time in the 1990s. This book would not have been possible without both of them. The book is also dedicated to my loving wife, Rebecca, who endured my extended reporting trips abroad and many late-night Skype calls to foreign lands, and who ultimately believed in me. I cannot thank her enough.

Sam Freedman, a true *mensch*, pushed me from the very beginning, when this book was just a sketch of an idea for his book writing class at the Columbia University Graduate School of Journalism. I am forever indebted to him and his mentorship, scholarship, and editing skills. I am lucky to consider him a colleague and friend.

My editors, Marlie Wasserman and Doreen Valentine, saw this book through its many iterations, helped shape it, and shepherded me through the process. Thanks also to Noah Breuer for providing cover art for much less than what he deserves.

A major shout out goes out to three good friends who edited the manuscript in its earlier stages: John Borland, Joe Lewandowski, and David Sasaki.

Big ups to family members who have always shaped my world-view: Alex Farivar, Zari Farivar, John Hadsell, Virginia Hadsell, Heidi Hadsell, Martin do Nascimento, and Nena do Nascimento.

Un grand merci also for the friendship and support of David Boyk, Rachel Rosmarin, Dallas Bluth, Michele Jonas, Joe Lewandowski, Angie Lewandowski, Monica Tirado, Malcolm Knapp, Julie Pinkerton, and Zach Pinkerton.

Also, *muchas gracias* to Alan Wiig for his never-ending enthusiasm for this project and for suggesting the title and Bhu Muhler for his inspirational photograph of an Internet café that I have kept on my desk for many years.

I am also grateful to my Columbia classmates and journalism colleagues, most notably Denise Carson, who commiserated with me over the minutiae of book writing. *Medasi* and *спасибо* to Ethan Zuckerman and Evgeny Morozov, respectively, for always challenging my thinking on global Internet issues.

This book also would not exist without its four major characters: Chon Kilnam, Amadou Top, Veljo Haamer, and Omid Memarian. Thank you for allowing me to spend so much time with you and to tell your stories.

IN THE SOUTH KOREA SECTION, 감사합니다 to Jim Larson, Park Huynje, Chris Chung, Bob Kahn, Florence Chee, Hanna Cho, Ted Peng, Yong Sam Koh, Hyon Oh Yoo, Matthew Weigand, Jin Ho Hur, Vint Cerf, Kwang Oh Suk, Park Seung-Kyu, Jung Chul, Paul Lee, Sunay Lee, Lee Yong-Teh, S. H. Kyong, Jake Song, Kim Yoon, Lee Dongman, Huhh Jun-Sok, Daniel Lee, Je Hun-Ho, Lee Yeon-Yeol, Jean Min, Mitchell Hong, M. K. Kang, Cheon Yeong-Cheoul, Kim Hyesoo, Carl Seaholm, Paul Chong, and Don Sutherland.

I traveled to report in South Korea in April 2007 and will be forever indebted to the CouchSurfers and new friends who opened up their homes to me: Karla Moore (thanks especially for your spare mobile phone!), Loren Everly, Rachel Stotts and Nathan Lucas, Jennifer Lee, Aaron Tassano and Shin Soo-Jin, and also to Gary Rector for a truly unforgettable evening.

IN THE SENEGAL SECTION, *jëre-jëf waaye* to Jim Delehanty, Olivier Sagna, Blaise Rodriguez, Daniel Annerose, Bamba Baba, Bob Kahn, Brian King, Cheikh Mdiouck, Daouda Mbaye, Erin Soto, Fatoumata Sow, George Sadowsky, Ibrahima Basse, Ibrahima Yock, Janine Firpo, Lisa Goldman, Emily Renard, Sylvia Lankford, Mohammed Tidiane Seck, Jeynaba Ba, John Mack, John Stamm, Judith Payne, Larry Landweber, Mamadou Gaye, Matt Berg, Max Gaye, Madame Diagne, Ousmane Diop, Ousmane Mbaye, Russell Southwood, Scott Pennington, Abdoul Ba, Alain Just Coly, Alioune Faye, Tiffany Lee Brown, Fatimata Sèye Sylla, Ben Akoh, Babacar Diop, Steve Huter, Cherif Sarr, Jean Pouly, Anne Koplinka-Loehr, Marian Zeitlin, and Alex Corenthin.

I traveled to Senegal in January 2007 and for three weeks experienced real Senegalese *teranga* courtesy of Fatou Lô Rochefort and Dominique "Issa" Rochefort.

IN THE ESTONIA SECTION, *aitäh* to Andrus Aaslaid, Ivar Tallo, Kristijan Kask, Siim Teller, Mart Thomson, Mihkel Tammet, Riihon Urksa, Allan Martinson, Rain Rannu, Madis Tiik, Sten Soosaar, Arne Kaasik, Oliver Wihler, Andrus Viirg, Raul Volter, Jacob Farkas, Liivi Kruus, Kris Haamer, Heiki Kübbar, Mart Laar, Madis Taupere, Enn Tyugu, Rainer Nõlvak, Meelis Kitsing, Andres Aarma, Tarvi Martens, Thad Hall, Sten Tamkivi, Taavet Hinrikus, Priit Alamäe, Andrew Vskimeister, Gunnar Kobin, Juhan Parts, Margus Kurm, Silver Meikar, General Johannes Kert, Kenneth Geers, Lauri Almann, Hillar Aarelaid, Heli Tirmaa-Klaar, Raul Allikivi, Anto Veldre, Ago Väärsi, Riina Reeder, Martin Jasko, Linnar Viik, Arvo Ott, Edmund Laugasson, Dawn Nafus, Rica Semjonova, Martin Leiger, Jaak Aaviksoo, Luukas Ilves, and President Toomas Hendrik Ilves.

Tuhat tänu for the couch, endless nights of pool, beer, hospitality, saunas, and camaraderie to Ivo Kivinurk, Vaino Haamer, Kris Haamer, and especially Veljo Haamer.

Tänan for the Estonian translation, Maria Visnapuu, and to Liene Lapevska, *paldies* for the drinks in Riga. Finally, *vielen dank* for the couch in Berlin, John Borland and Aimee Male.

IN THE IRAN SECTION, ممنونکرم خیلی to Aaron Scullion, Abbas Milani, Afshin Memarian, Afshin Molavi, Ali Shokoufandeh, Behdad Esfabod, Borzou Daragahi, Bozorgmehr Sharafedin, Dariush Zahedi, Hamid Tehrani, Farnaz Ghazizadeh, Hadi Ghaemi, Hanif Mazroi, John Kelly, Jonathan Lundqvist, Leila Memarian, Mahan Abedin, Mark Glaser, Massoud Safiri, Mohammad Ali Abtahi, Nasrin Sotoodeh, Nima Mina, Parastoo Dokouhaki, Payman Arabshahi, Ramin Jahanbegloo, Roozbeh Mirebrahimi, Salman Jariri, Shahram Rafizadeh, Shahram Sharif, Shideh Bahmanyar, Siavash Shahshahani, Sina Motalebi, Sina Tabesh, Steve Goldstein, Touran Memarian, Afshin Memarian, Sanam Dolatshahi, Hossein Derakhshan, Hamed Derakhshan, Ali Reza Eshraghi, Farhad Ardalan, Babak Siavoshy, and Austin Heap.

Mersi to Rebekah Kouy-Ghadosh for the Persian translation and your unwavering friendship. Finally, an extra special *ma'am noon* goes out to my cousin and Iran analyst extraordinaire, Karim Sadjadpour, for introducing me to Omid Memarian.

The Internet of Elsewhere

Introduction

Tallinn, Estonia

June 19, 2009

One week after the Iranian presidential election I found myself sitting alone at my friend Veljo Haamer's desk. I scoured the Internet for information about Iran. The country had been essentially at a standstill and the foreign media expelled as the government clamped down on what little free speech remained. Occasionally, I sipped from my Estonian lager and stared out his second-story window. I watched the hours-long summer twilight burning through Haamer's curtainless east-facing windows. On that night, I was alone, as Haamer had gone to visit his father in the eastern countryside, near the Russian border.

Veljo Haamer is Estonia's well-known WiFi evangelist and had become a good friend ever since my first trip to Estonia in March 2005. On each of my next four trips to Tallinn, I'd slept on his couch. The studio apartment wasn't very big at all. In fact, it was just a single room that'd been divided into roughly four quadrants through the creative use of furniture. One housed his bed (an inflatable air mattress), his Apple AirPort Extreme WiFi router, his desk, and a bookcase. Towering above his workspace along a wide bookcase was a mishmash of wireless cards, WiFi-related books, magazines, old ticket stubs, and flyers, and his own WiFi.ee stickers strewn about in a very haphazard fashion. Old conference badges flopped all over random business cards cozying up to back issues of *Wired*. On the other side of this bookcase was the tiny second quadrant, with only a rarely used television and a dusty couch.

A stand-alone closet served as a barrier between the couch and the kitchen, creating the third quadrant, which had barely enough room for one person to cook at a time. On this trip, Haamer's fridge was finally stocked with more than just beer, sausage, butter, and bread—the first time I'd seen anything of substance in it since I'd known him. He admitted to me that the previous six months had been especially difficult for him. His freelance WiFi consulting work had all but dried up. That meant not eating or drinking out nearly as much as before, and not driving at all. Haamer was too proud to go on public assistance, saying that he should "earn incomings by my own hands."[1] The final quadrant, the entryway and bathroom, had a shower stall that was so close to the kitchen sink that anyone taking a shower could easily reach for the dish soap above the sink just by sticking a hand out of the shower door. The toilet was the only other room in the apartment with a separate door, and even then it was just big enough to hold a toilet, a shelf, and a poster of a German band, Rammstein.

Haamer's apartment is one of eight units in a small building at the end of Leigeri Street in Kalamaja, one of the Estonian capital's oldest suburbs. Local lore has it that Kalamaja dates back to the fourteenth century, when it was a mere fishing village. Today, it's a charming neighborhood filled with small wooden houses from the 1920s. It's one of the few areas in Tallinn that was not damaged during the Soviet occupation of Estonia, which lasted from 1945 until 1991. Although it's just a short walk from Tallinn's train station, and a little farther to the Estonian Parliament, it feels like a world away from the increasingly expensive, fast-paced, cosmopolitan, English-speaking city center where many foreigners spend much of their time.

Officially, I had been in Estonia to speak at the first NATO Cooperative Cyber Defence Centre of Excellence conference on cyberwarfare and cybersecurity. But at that late hour, I was working on a story for Public Radio International's *The World* about how Silicon Valley companies, including Apple, Google, and Facebook, had just released new Persian-language capabilities in the wake of the controversial Iranian election. Enthroned in Haamer's chair, I used Skype to place free calls over the Internet to sources in California. Skype had been created just a few years earlier, only a few kilometers from where I sat in Kalamaja. Here I was bringing together California and Iran through Estonian engineering. In just a few short hours, I conducted interviews, wrote, edited, and recorded my entire piece for a nationally syndicated American radio program. While I made a late dinner in the midnight twilight, I listened to the show live over the Internet. Less than a decade earlier, this would have been impossible. These were just the latest iterations of new technology tools that have become vital to Iranian dissidents, given that all independent foreign media had essentially been kicked out of the country, and that free domestic media did not exist.

Weeks before the election, I had reported on the increased use of social media, including Twitter, Facebook, and FriendFeed by all major candidates in the Iranian presidential election, reformist and conservative alike. Even President Mahmoud Ahmadinejad himself had started a blog back in 2006. But in an environment with almost no nongovernmental voices, and a crackdown on the existing foreign satellite broadcasts, reformists and protestors had no choice but to turn to the Internet, to send e-mails, photos, and video. One Persian-language satellite television channel based in southern California was said to have sent thousands of miniature pen cameras to Iran, which could easily be connected to a computer to upload their photos to the Internet. Reformist candidates, most notably Mir-Hossein Mousavi, the former prime minister, used his Facebook page on June 13 to call for supporters in Tehran to go to their rooftops and shout the normally religious chant "Allah wa Akbar!" (God is Great!), thereby co-opting it from the hard-liner religious conservatives. He wrote: "Tonight all go to the roofs and let the chant, 'Allah Akbar' fill the air in Tehran. Internet and cell phones are completely offline. Use landlines as much as you can and spread the news."[2]

Later, messages in English and Persian spread quickly across Twitter, with a small handful of users in Iran sending out as much information as possible. Sina Tabesh, a twenty-three-year-old Internet user in North Tehran, tweeted in English on June 13, 2009: "Northern part of Tehran is on fire, people attacked a gas station to explode it." Others were spreading information of arrests of dissidents as the government disseminated it: on June 13, 2009, another blogger, Somayeh Tohidlou, wrote on her FriendFeed page in Persian: "Latest News: Doctor Mohsen Mirdamadi—Ms. Zahra Soon Mirdamadi wife—Saeed Shariati—Zohreh Aghajari and Behzad Nabavi, Abdollah Ramezanzadeh are confirmed arrested."[3]

While I waited for my piece to be edited and mixed, I ran Twitter searches, looking for the latest news and information about Iran, checking messages tagged *#iranelection,* a searchable code word to find more information about the Iranian election. Many journalists, myself included, felt overwhelmed with the sheer volume of Twitter messages pertaining to Iran. The torrent of messages providing links to photos and videos, watching them as they came in—particularly when more than usual claimed to be coming from within Iran—was unlike anything I'd ever seen. The best and most newsworthy messages, I forwarded on to friends and sent out on my own Twitter account.

Some messages included raw emotions, like this one from Parastoo Dokouhaki, a young feminist blogger in Tehran, who wrote in English: "[I] feel like somebody's foot is on my neck, preventing me to breathe. I can do almost nothing online. Suppression. Want to cry. #IranElection." Others expressed frustration at the seeming lack of help by foreign tech companies, like Google. As Babak Mehrabani wrote in Persian: "A few days ago [companies made] any excuse not to service the Iranians, now special services for Iranians are starting? Example: Google Translate."[4]

THE MORE I READ THESE POSTS, the more I felt like I was part of something bigger than myself. With information flowing faster than I could read it, I got swept up in the fervor, as did many others across the globe. But with my own Iranian-American heritage, and my extended family in Iran, how could I not try to understand the play-by-play of actions and analysis that were coming out about Iran? This time, though, many non-Iranians and others who had no previous connection to Iran somehow suddenly became captivated as well. Many users changed their Twitter avatar color to green, in solidarity with the opposition campaign of Mir-Hossein Mousavi. The former prime minister's campaign used the traditionally Islamic color of green as a way to co-opt the religiosity of his conservative opponent, President Mahmoud Ahmadinejad. I also noticed that many Twitter users had changed their virtual location to Tehran, hoping to confuse Iran's Internet police—the idea being that if everyone is in Tehran, then no one is.[5] There's no evidence that this strategy worked, but that didn't stop many people from trying to help, even in some small way.

Indeed, later research by the Web Ecology Project, an academic interdisciplinary research group based in Boston, confirmed that between June 7, 2009, and June 26, 2009, there were over 2 million messages on Twitter about the election in Iran. While it was easy to think that these Twitter messages were part of a mass movement, the top 10 percent of users produced two-thirds of the Iran-related tweets, and one-quarter of all tweets about Iran were retweets, or rebroadcasts, of someone else's messages.[6] This research confirms Twitter's role as more of an amplification and dissemination tool. It was not primarily used as a social organization tool from within Iran.

However, while relatively few Iranians were using Twitter prior to the election, nearly all those who had a mobile phone—roughly two out of every three Iranians—were familiar with text messages. As is the case in many parts of the world, text messages are the cheapest and most effective way to communicate quick pieces of information in Iran. The Islamic Republic clearly knew that mobile phones would also be used as a way to socially mobilize dissidents, and blocked all text messages on all three carriers from Election Day until July 1, 2009.[7]

EARLIER THAT WEEK, on June 15, 2009, the U.S. Department of State realized the potential power that Twitter could have in distributing information out of Iran. A State Department official reached out to a Twitter founder and suggested that the company reschedule its planned maintenance time—during which the entire service would be shut down—so that it would be more conducive to users in Iran, instead of American users. Twitter complied, noting at the time that the service "is currently playing as an important communication tool in Iran."[8]

Even before the State Department took notice of Twitter, there were other forms of protest that many reformist sympathizers abroad began to take into their own hands. These activists targeted Iranian government websites, including that of President Mahmoud Ahmadinejad, as a way to show disapproval. This tactic, known as a "denial of service" attack, uses false traffic to overwhelm a web server, causing it to crash. It would be as if a highway could suddenly get flooded with so many cars that no car could possibly move.

Although I didn't know it at the time, tech-savvy activists like Austin Heap, a twenty-five-year-old "cyberactivist" in San Francisco, wrote scripts and created new sets of instructions that specifically targeted Iranian government websites. Non-techies discovered existing tools, like PageReboot, designed for commercial uses like monitoring auctions on eBay, and configured them to automatically reload particular sites every few seconds. While some Islamic Republic sites did go down at various points in the immediate aftermath of the election, it's hard to say how effective the calls for such cyberattacks were. Nonetheless, countless anonymous Internet users worldwide felt compelled to show their defiance against the Islamic Republic.

Other online pundits provided a more sobering analysis of the effectiveness of such attacks, noting that because Iran had throttled down the overall speed of its domestic Internet by 60 percent (from five gigabits per second before the election to two), the cyberattacks made it that much more difficult for pro-opposition websites to get their message out. James Cowie, the CEO of Renesys, a bandwidth analysis firm, put it this way in an e-mail to technology journalist Evgeny Morozov: For all the well-intentioned efforts to attack government websites in Iran, dissidents and their supporters may have been in fact just shooting themselves in the foot. He concluded that, technically speaking, "If you attack a pro-government site, you are almost certainly also stealing bandwidth from pro-opposition sites."[9]

The Islamic Republic and its supporters began reacting as well. Just as reformists and other dissidents were active on Twitter, so too were voices of the regime, including the supreme leader, Ayatollah Ali Khamenei. A Twitter account in his name—@khamenei_ir—linked back to his official website. There were also reports that government agents or supporters were using Twitter to spread false information, such as sending protestors to nonexistent protests.[10]

These actions by the government and/or their sympathizers mimicked the behavior of a cadre of reformist bloggers who had been active for years, writing from both within and outside of Iran about the failings, abuses, and crimes conducted by the Islamic Republic. The government, in turn, responded with a clampdown, and before the June election it announced that it would be launching an online army of 10,000 *Basij*—a morality police—who would act as pro-regime bloggers.[11]

Further, in the months leading up to the Iranian presidential election of June 2009, a government-run site, gerdab.ir ("whirlpool" or "vortex"), appeared and after the election posted photographs of protestors, asking the public to identify them.[12] If any government sympathizers have come forward with "useful" information, the Iranian government hasn't said so. This battle has continued on in subsequent months, with such headlines on the gerdab.ir website as "Iranian Revolutionary Guards ready to fight cyber and Internet war" (September 7, 2009) and "An Internet battle report in the defeated velvet coup" (October 1, 2009).

Clearly, the Islamic Republic is trying to make an online show of force—to underscore the point that it is capable in engaging online dissidents in its own way.

MEANWHILE, Austin Heap moved on from making offensive tools against the Islamic Republic to publishing a list of open proxy servers, or ways that Internet users in Iran could filter their traffic through computers outside of the country, thereby circumventing the government filtering system. But that effort was quickly overwhelmed as Iranians warned Heap that he was exposing those servers to attacks from Iran. Then he created a password-protected list of proxies and authored instructions to create more proxy servers.[13] That would let anyone online who was sympathetic to reformist Iranians create tools that would allow Internet traffic from Iran to be re-routed to computers outside the country, thereby establishing a crucial link between dissident Iranians and the outside world. Heap had no prior connection, professional nor personal, to Iran. But one of his core beliefs is that the Internet should be accessible to all. Period.

"Three weeks ago I was very happy playing Warcraft and I was following the Iran election," he told me in an interview in July 2009.[14] "But it wasn't until everything escalated there that I got involved." Heap felt that, given his technical skill and his belief in helping those being oppressed by their government online, it was his "responsibility" to help those less fortunate than himself.

Despite a quick outpouring of support from Iranians worldwide, the cyber authorities in Tehran were not having any of it. Within a few short days, Iranian authorities reacted to block all nonencrypted proxies—making the efforts of Heap, and everyone else who had followed his instructions, essentially useless. While Heap continued to probe the ability of Tehran's cyber authorities, other existing online anti-filtering tools began to spread around the Internet as well. One of the oldest (and arguably the best) of these tools was Tor. It was created in 2003 and was initially funded by the U.S. Naval Research Laboratory, and later by the Electronic Frontier Foundation. Today, its parent organization, The Tor Project, is an independent nonprofit based in Massachusetts.

Tor was initially designed as a web-anonymizing program, making it possible for people to surf the Internet as anonymously as possible. Tor's creators discovered that their software was also able to circumvent anti-filtering programs, even though it was not expressly designed to do so. As the most sophisticated of many anti-filtering applications, Tor strips away online identifying information as data passes through its network. That means it's significantly more difficult for anyone, including Iranian Internet authorities, to surveil online activity.

Tor relies on a network of thousands of users around the globe who have installed the free software on their computers and have them set up to act as relays. Once a computer is configured to act as a relay, other Tor users can pass data through it. Each time a Tor user accesses the web, Tor passes the data through different, constantly changing relays as a way of masking the user's online trail.[15] Word of Tor and how to use it spread across the Iranian Internet very quickly.

"Before the election we were seeing about one to two hundred new users [from Iran] per day," said Andrew Lewman, executive director of The Tor Project, in an interview on July 2, 2009. "Right after the election and as the protests started, we started seeing that spike up into seven hundred to a thousand per day. Now we're up to about two thousand new users a day and around eight thousand connections sustained at any time, which is a huge, dramatic increase."[16]

Another anti-filtering tool that received a surge of attention was Psiphon. This application was released in early 2007 as a spin-off project from the Citizen Lab at the University of Toronto. However, less than ten days after the election, Psiphon's chief executive officer, Rafal Rohozinski, told the *Ottawa Citizen* that his colleagues were seeing one new Iranian user on the Psiphon network every minute. By the end of June, Rohozinski and his colleague Ron Deibert reported that more than fifteen thousand Iranians had used Psiphon. Again, Tehran took notice. As a result of Psiphon's popularity, Rohozinski and Deibert wrote, the Canadian envoy to Iran was "admonished" for helping Iranians get around government filters and access "immoral" content.[17]

One of Psiphon's attractive qualities is that it does not require any special configuration or the downloading of any additional software. In fact, since the Iranian election, Psiphon has released a special webpage in Persian for Iranians to sign up and use the service without having to download or install any software.[18]

"Psiphon is good for non-techies, easy to use types who don't want/know how to install software on their computer," wrote Nart Villeneuve, Psiphon's chief technology officer, in an e-mail.[19] "Tor is best for security, people at risk, those that require security."

WHILE SOME IRANIANS WERE TURNING to Tor and Psiphon for help, Austin Heap wasn't giving up. He tightened his circle and began working with a group of programmers and activists around Europe, Iran, and the United States, including Daniel Colascione, a similarly minded cyber activist living in New York State. On July 4, 2009, Heap announced the launch of Haystack ("Good luck finding that needle"), a new program specifically designed to target the Iranian Internet filtering system. Relying on information leaked from within Data Communications of Iran, Heap and others went to work on creating a computer program that they claimed could fool the Iranian Internet surveillance system.

Within days of the announcement, Heap hastily convened an evening meeting in a San Francisco high-rise with Colascione and a group of Iranian-American programmers, attorneys, and business leaders. By this time, I had returned to California from Estonia and joined this group. There, Heap demonstrated Haystack for the first time and discussed the possibility of sneaking the program into Iran on smuggled USB sticks, digital camera memory cards, and other small, inconspicuous forms of media. Unlike Tor and Psiphon, the plan was not to make the application publicly available, but rather to have it be slowly distributed across a network of trusted users who would have corresponding numerical codes. These codes were put in place so that if Haystack fell into the wrong hands, then that portion of the network could be disabled without risking exposure of all users.

As the summer progressed, Heap continued to work on Haystack and got the attention of Washington, D.C., insiders. He met with aides, political analysts, and various congressional leaders, including Speaker of the House Nancy Pelosi, to discuss the possibility of getting financial and organizational support from the federal government for Haystack. Meanwhile, he continued improving the software. It wasn't until April 14, 2010, that Heap announced that Haystack had received a license for export to Iran from the U.S. Department of the Treasury. However, by September 2010, Haystack fell apart due to mistrust amongst its creators and the fact that a leaked copy of an earlier test version was exposed by security researchers to have serious flaws. Colascione and the board of directors immediately resigned, and the Haystack website stated that the organization had halted testing pending a security review.

BUT DESPITE THE UNPRECEDENTED LEVEL of attention given to Iran from both inside and outside the country, and Heap's advancing political connections, the status quo was preserved. Amid international outcry and violent repression at home, President Mahmoud Ahmadinejad was inaugurated for a second term on August 5, 2009. If there were any instance when an online community should be able to "fight the power," it should have been in the summer of 2009. After all, Iran is a country with a young, literate, and highly wired population. Given its history of political turbulence, coupled with the new social networking capabilities, online "netizens" might be expected to influence local politics in a meaningful way.

However, Iran has also illustrated its capability and willingness to limit Internet and text messaging access, to infiltrate online communities, and to intimidate dissidents. Indeed, there is evidence to suggest that the incumbent regime is being consolidated through military and political power, despite all of the international efforts from inside and outside of Iran to speak up online. But the Islamic Republic is constantly playing catch-up, as dissidents get their message out any way that they can. Less than two months after Ahmadinejad's swearing-in, on September 27, 2009, Mobin, a company entirely run by the Islamic Republic's Revolutionary Guards, bought a controlling stake (50 percent plus one share) in the Telecommunication Company of Iran. It paid $7.8 billion for the privilege—the largest single deal in the history of the Tehran Stock Exchange.[20] This acquisition further suggests that the regime's desire to control online access and activities within the country. Despite this new level of control, blog posts, photos, and videos continue to trickle out of the country to this day.

In the late twentieth and early twenty-first centuries, conventional wisdom dictated a technological determinist view, that the mere presence of such a revolutionary telecommunications technology would upend the existing political, economic, and social regime. MIT's Nicholas Negroponte famously argued in 1995 that "being digital" could "flatten organizations, globalize society, decentralize control and help harmonize people" and that "overly hierarchical and status-conscious societies will erode. The nation-state may go away." He added: "Developing nations will leapfrog the telecommunications infrastructures of the First World and become more wired (and wireless)."[21]

On its face, this may not be such a ridiculous notion. Conventional social science theory argues that interpersonal relationships can beget community, which can beget widespread solidarity, which can beget enlarging power, which can beget greater authority, which can finally beget revolution. However, a quick check of history and the Internet's historical analogy—the telegraph— shows us that this theory does not quite hold up.

In the nineteenth and early twentieth centuries, the telegraph promised to connect the world as never before. People around the globe marveled that this new technology could now move information with incomparable speed. As Captain George O. Squier of the U.S. Army Signal Corps wrote in the January 1901 issue of *National Geographic*: "The fastest mail express, or the swiftest ocean ship, are as naught compared with the velocity of the electrical impulse which annihilates any terrestrial dimension."[22]

As the telegraph began to spread to all regions of the globe, many intellectuals like Henry Field (a Presbyterian pastor, author, and brother of Cyrus Field, who laid the first direct telegraph cable between Europe and the United States in 1858) began to posit that perhaps this global network could even bring about world peace. He wrote shortly after the trans-Atlantic cable was laid that the telegraph "unites distant nations, making them feel that they are members of one great family. . . . By such strong ties does it tend to bind the human race in unity, peace, and concord."[23] Nearly a century later, similar intellectuals marveled at the astonishing power of the Internet, and its best-known application, the World Wide Web.

In a 1993 article about the nascent web, the *New York Times*'s John Markoff wrote with much of the same tone of marvel that Squier did at the beginning of the century:

Click the mouse: there's a NASA weather movie taken from a satellite high over the Pacific Ocean. A few more clicks, and one is reading a speech by President Clinton, as digitally stored at the University of Missouri. Click-click: a sampler of digital music recordings as compiled by MTV. Click again, et voila: a small digital snapshot reveals whether a certain coffee pot in a computer science laboratory at Cambridge University in England is empty or full.[24]

Perhaps this gee-whiz utopian notion may have worked in the early days of both of these technologies, but as the Internet has spread around the globe, it has not brought a single country that much closer to "harmonizing people"— indeed, crime, conflict, and war all still exist, just as they did at the peak of the telegraph-connected world.

WHAT HAS BEEN HAPPENING in Iran during the two decades since the Internet first arrived is far more interesting than the simple narrative of a young wired generation throwing off the yoke of an oppressive regime. Iran is a key example of a more complicated and protracted struggle between a government and its people online, a constant see-saw situation in which each side can have the upper hand, depending on the week, or sometimes even the day. It's more than just a tale of societal progress and democratization as brought about by the Internet. It's deceptively simple to proclaim that Iran's online voices are "nothing short of a revolution within the [Islamic] Revolution," as the Iranian-British author Nasrin Alavi (a pseudonym), wrote in her 2005 book, We Are Iran.[25] Indeed, for every voice yearning for freedom and reformist politics, there are other voices that simply choose not to engage, voices that are silenced by intimidation and threats, and still other voices that are made more cautious by government agents infiltrating or spreading disinformation on social networks.

It is important to recognize the political, economic, and social context in which Iran's unique turmoil online has happened. Iranian politicians did not use Twitter and Facebook by accident. Iconic videos of protestors—and sometimes their tragic deaths—were not disseminated on YouTube randomly.

Iran has a unique political situation as an Islamic Republic that attempts to be semi-democratic within the confines of an authoritarian Islamic regime. In Iran's previous elections, subsequent to the establishment of the Republic in 1980, there were disagreements and even protests. Many younger citizens were frustrated that President Mohammad Khatami's social reforms of the early 2000s did not go far enough to allow for more newspapers and other forms of political freedom. These "reformists" were not calling for an out-and-out overthrow of the Islamic Republic, merely substantial reforms within the larger structure. However, in 2009, large numbers of men, women, and children gathered together in the streets of Iran's biggest cities to fiercely oppose what they believed to be outright fraud in an election that the government proclaimed to be authentically democratic. In other words, the government couldn't even win an election that was rigged to begin with. After all, it had disqualified many candidates who did not meet the adequate "religious" qualifications.

From an economic standpoint, Iran has been in shambles. Since Ahmadinejad took office, unemployment has been estimated to be as high as 20 percent. Further, 80 percent of Iran's state revenue is taken from oil and gas sales, and the significant drop in oil and gas prices since 2007 has caused that revenue to drop precipitously. Further, President Ahmadinejad has not slackened his economic handouts, low-interest loans, and lots of state spending—leading the country further into debt. Worse still, inflation hovers at 25 percent, with no signs of abating.[26]

These economic troubles are hitting Iran's youth especially hard—this is, after all, a country where half of the population is under twenty-five years old. These "Children of the Revolution" were born after 1979, during a time when Ayatollah Khomeini encouraged Iranian women to have more children while many men were being killed during the 1980–1988 Iran-Iraq war. But now, these children are growing up in a country that can educate them—it was more than ten years ago that women began to outnumber men in Iranian universities— and they are finding that their country cannot adequately provide work and social services for them. So, whether the Islamic Republic likes it or not, an entirely new, educated, and wired generation has grown up in Iran with the Internet, bootleg satellite television, and mobile phones. However, in a country that cannot create enough jobs for them, in a country that restricts sports, music, and other social freedoms (particularly between the sexes), it's no wonder that so many go online as a form of cheap entertainment.[27]

It is this environment of international and domestic political turmoil, economic difficulties, and social restrictions that drives overly educated and under-stimulated Iranians to the Internet. This confluence of events has created a completely unique online outcome that's far more interesting and unpredictable than the Negroponte-style techno-determinism. It's no wonder that Iran was the first country to arrest a blogger (2003), and sadly, the first country to have a blogger die while in prison (2009). It's no wonder that Iran has been active in trying to influence domestic online activity for over a decade, and has engaged in one of the world's most complicated and devious Internet filtration systems, second only to China. It's no wonder that in the aftermath of the June 2009 election the Internet became a key tool in getting information in and out of Iran, as foreign media were barred from the country and the country's own media remained highly restricted. In other words, as the Internet has collided with Iran's political, economic, and social realities, it has created a conflict far more complicated than anyone could have imagined.

JUST LIKE IN IRAN, when the Internet arrives in any country, it bumps up against various preexisting political, economic, social, and cultural histories and contexts—and often what comes out are rather surprising results. The most fascinating examples of Internet-related changes are not happening in Silicon Valley, but rather in far-off, forgotten, or overlooked corners of the globe.

Most Americans are generally familiar with the rapid rise in Internet technology in the 1990s as more and more people became connected with faster and faster connections. Many of us remember the go-go 1990s, when many companies raised unimaginable sums of venture capital on the thinnest of business plans, only to come crashing down in the early 2000s. We have become accustomed to hearing about California companies like Google, Apple, Facebook, Twitter, and many others that purport to "revolutionize" the way we gather, store, and share information.

However, in the United States, we often forget about the Minitel—first released in 1982—France's predecessor to the Internet, which allowed users to check sports scores and buy train tickets through the computer terminals in nearly every home. We're largely unaware of the fact that across the European Union most banks can easily accept online electronic bank transfers for free, while most Americans still pay with paper checks—a technology that has not changed substantially since the early twentieth century.

Looking at Asia, we often don't know about how China, in many ways, could represent the future of the Internet, in that it has the largest single user group from one country, and yet it has the most sophisticated filtration and censorship system in the world. In South Asia, we celebrate the rapid rise of India and its middle class-staffed call centers as the center of Thomas Friedman's "flat world," without understanding how or why this nation used its education system and English-language ability to leap forward as a result of its rich modern history.

Turning to Latin America, few Americans know the story of how in 1996 Brazil became the first country in the world to introduce electronic voting, more than four years before the Help America Vote Act that emerged in the aftermath of the November 2000 American election debacle. Many of us also forget about the meteoric rise of StarMedia, a Spanish- and Portuguese-language web portal founded by a thirty-year-old Uruguayan who grew up in Connecticut. When StarMedia went public in 1999—the first Latin American Internet company to do so—it hit a market capitalization of over $1 billion, later reaching $3.8 billion.

Africa, of course, is still stereotyped as the "dark continent" even well into the twenty-first century, as it lags behind in electricity production and Internet connectivity. However, South Africa is regarded as a contemporary beacon of economic growth and hope for the entire continent, especially after having hosted the World Cup international soccer tournament in 2010. After all, it has the highest gross domestic product in all of Africa, and has the highest rate of Internet connectivity, with approximately 11 percent of its population online.

Furthermore, the African National Congress was one of the earliest African political parties to maintain a web presence. However, while much of Africa is under-connected to the Internet, and areas of East Asia and Northern Europe have the highest rates of Internet penetration, we must ask ourselves how exactly this online landscape has come about, and why certain applications—ranging from Kenya's M-Pesa mobile phone and online payment system to Canada's ubiquitous e-mail device, the BlackBerry—have emerged from their respective countries.

We are aware of how the Internet is fundamentally a means to transmit individual, discrete ideas from one place to another, but we are often not fully cognizant of how the Internet can transmit the collective spirit of a nation as well. When the Internet arrives in all countries, it does so at a particular moment in history, and evolves in a way that is irrevocably stamped with those countries' modern histories and economic environments. The net result is that all countries' online applications and cultures are inevitably distinct, with differences derived from very local characteristics, such as the quality of middle-class education, or the culture's particular stripe of national drive.

But while every nation in the world has a unique story with respect to its Internet history, I have chosen to examine and tell the Internet history of Iran, with its story of what happens when an Internet-fueled younger generation clashes with an authoritarian regime. I will compare Iran's Internet narrative to the tale of three more of the world's most technologically fascinating countries: South Korea, Estonia, and Senegal, one of sub-Saharan Africa's best-connected nations. These four nations display the broad range of their contributions to the Internet along the world's continuum of connectivity, from most to least wired.

While nearly every cultural distinction leaves its own imprint on any nation's Internet history, several characteristics appear to exert an outsized effect. Education, and most particularly the most basic literacy that allows potential users to read, write, and otherwise interact with webpages, is a critical variable. The citizens of highly educated nations, like South Korea, Estonia, and Iran, have found little difficulty in translating ordinary literacy into Internet literacy. Nations like Senegal, where literacy rates are relatively low (40 percent), have a much higher hurdle to overcome in order to achieve widespread Internet usage.

A particular national pride that manifests as a do-or-die independent spirit often plays a role as well—indeed, Estonia and South Korea both blossomed digitally after dark spots in their history, and locals see their online prowess as an expression of national resiliency and pride. Both countries, in fact, have tied their contemporary identity to being exemplars of technological modernity through urban design, and the ubiquity of wireless communications tools. In Iran, a nation that has a particularly long and proud history, the struggle for the country's soul is borne out online. Many young, well-educated, secular Iranians use the Internet as the only free public space that is available to them, and struggle to stay one step ahead of the authoritarian Islamic Republic, which is trying to control this new medium as it has television, radio, and newspapers before it. Senegal, meanwhile, has attempted to use grandiose government-led visions through governmental plans, presidential directives, and the efforts of outside organizations to become a leader for West Africa and the rest of the subcontinent, and yet it remains stymied at every turn.

South Korea, the world's most wired country, anchors one end of this continuum. Estonia is a tiny country that, despite having been occupied by nearly all of its neighbors in its history, emerged from nearly fifty years of Soviet rule and in less than twenty years created the world's first national digital ID card system, online voting in national elections, and Skype. Senegal represents the best of what sub-Saharan Africa has been able to do, and yet this West African nation still falls significantly short of what the continent will need to advance further in the twenty-first century, as it has been unable to translate the efforts of a few motivated, well-educated individuals into connectivity for a less-advantaged population.

Despite the fact that broadband Internet access began in the United States in the 1990s, South Korea is now the world's leader in high-speed connectivity. In fact, before half the American population had broadband in their homes, South Korea had a higher percentage of its people using the Internet many times faster. On May 5, 2003, the *New York Times* ran this headline: "America's Broadband Dream Is Alive in Korea." The article reported that South Korea had "built the world's most comprehensive Internet network, supplying affordable and reliable access that far surpasses what is available in the United States, even in those homes that have their own broadband setup."[28] This high level of connectivity resulted in unprecedented new forms of politics, media, and entertainment—indeed, Korea maintains the world's most highly developed professional computer game league. Its highest-paid player makes $300,000 per year, simply for clicking and typing, albeit at lightning speed. This raises a host of questions: How did a country decimated by war in 1953 multiply its gross national product per capita a hundredfold half a century later? How did a country with an obscure data-networking project in 1982 explode to having high-speed Internet in every home twenty years later?

Estonia is nothing short of the most exciting country that few tech-savvy Americans have ever heard of. Despite the fact that WiFi was pioneered in the United States, Estonia has taken this technology to new heights. Through the efforts of one man, Veljo Haamer, WiFi has become a near-ubiquitous tool available in all corners of this small country, almost always for free. While WiFi often costs several dollars or euros in the West, in Estonia it remains largely free, even in locations as unlikely as commuter trains, ferries, buses, countless parks, bars, and cafés. Further, while America is still unable to get paper-based voting to work in a flawless manner, Estonia has moved on to widespread Internet-based voting—the first country in the world to undertake such an endeavor. How did this tiny, forgotten country, once a corner of the Soviet Union, become the home to Skype?

While the Internet has helped countries in the global North (like South Korea) to zoom ahead of their relative competitors, this has not been the case in the world's most under-developed region, sub-Saharan Africa. Despite the fact that many technologists, political scientists, and theorists hoped that African countries would be able to "leapfrog" into the twenty-first century, this expectation has largely not been realized. Senegal is one of a handful of countries where the Internet stands a reasonable chance of having a meaning-ful impact on local people's lives. It is a politically and economically stable country that has the second highest level of bandwidth on the continent, behind only South Africa. However, the Internet in Senegal, while of relatively high quality (broadband connections are available), remains woefully behind; prices continue to be artificially high, and only around 10 percent of the country's population is online. Why hasn't the Internet taken off, despite well-intentioned and well-funded efforts from the United States and within Senegal? If the Internet cannot work in one of the continent's most promising locations, what hope does the rest of Africa have?

The Internet is not, in fact, a seed. It does not have the ability to bring about world peace and the elimination of the nation-state, any more than the telegraph did. It is but a tool that, when combined effectively with local politi-cal and economic realities, can have demonstrably positive and often surpris-ing effects. However, this tool can be co-opted and/or fought against by regimes that are not ready for it to be used freely. Other developing societies, too, may not be completely ready to use the Internet effectively. This is why manifesta-tions of the Internet remain so varied in different corners of the globe. This book is an attempt to tell the story of what happens when the Internet collides, head-on, with history unfamiliar to most Americans. This is the tale of the Internet of elsewhere.

1

South Korea

National power does not come from guns, but from the accumulation of science, technology, and information.

−Kim Young-sam, president of South Korea (1993–1998)

The Battle

Seoul, South Korea

April 16, 2007

The Yongsan district is a nexus of modern Seoul. It has a high-speed rail stop, a major subway station, the older Yongsan Electronics Market across the tracks, and its newer cousin inside the station's shopping mall. Here, one can buy anything that plugs in. However, the most unique feature of Yongsan is perched on the seventh floor of the expansive I'Park Mall, adjacent to the electronics market. There, all by itself, sits the eSports Stadium. The rooftop patio between the elevator and the entrance to the stadium is cold, plain, and quiet. There are none of the trappings normally associated with a sports stadium—no vendors, tailgaters, or rowdy fans.

The "stadium" is actually a room about the size of a recital hall, but instead of music, the stage is devoted to video games—specifically, StarCraft. First released in the United States in 1998, this game has faded from view in the decade since its release. While most kids in other parts of the world have moved on to newer games, like Counter-Strike or World of Warcraft, in South Korea, StarCraft reigns supreme. Although StarCraft II was not released until mid-2010—it was first formally announced at an event in Seoul in May 2007—the existing version had not changed in over ten years.

At the eSports Stadium, professional matches for a twelve-team league take place, complete with coaches, uniforms, referees, stage lights, theme music, smoke machines, television crews, and play-by-play announcers. On this day, around sixty spectators gather to watch their favorite "sport." The majority of Korea's adults don't quite understand the appeal of these games, nor do they understand why their children sometimes spend hours upon hours enraptured in this fantasy world.[1]

The crowd on this day consists mostly of groups of high school girls, sitting together and hiding their faces behind banners that they've brought, so they can cheer on their favorite players without being caught by television cameras. There are no concessions, nor official T-shirts—only these homemade signs. This match is pure sport and no marketing.

Like every Pro League match since its founding in 2002, this match is broadcast live throughout the country. Two television channels carry StarCraft twenty-four hours a day, with the larger one, Ongamenet, reaching 3 to 4 million viewers during the 6–10 P.M. prime time slot. If adjusted for population, the network's StarCraft broadcast in Korea would consistently top the weekly Nielsen television ratings in the United States. Most of the girls in the audience are supposed to be in school, or an after-school learning academy, or should at least be studying on a weekday afternoon, and definitely not sneaking out to watch competitive video games.[2]

As the lights dim and the smoke machine glazes the room with its translucent haze for dramatic effect, each team of seven players walks out onto their respective side of the stage. One of the teams, eSTRO, wears black-and-white tracksuits, with orange accents and a large eSTRO logo (written in English) on the left breast and shoulders, with a lightning bolt replacing the S. After a round of applause, cheers, and thumping techno music, the players calmly sit at a long counter.

Despite the incessant bass and stagecraft, the players don't appear to be excited. Not much more than a few smiles flash across the faces of the young men as they take their seats—no high-fiving, no trash-talking. It's all discipline, as if they were about to perform a work of classical music. As they enter the room, they proceed to sit quietly, seven in a row. The team's coach, Daniel Lee, anchors the team at one end; he's the only one not wearing a uniform, opting instead for a business suit. Waiting for the starting seven at the counter is a box of Dunkin' Donuts and a small bottle of orange juice at each chair. They are about to engage their opponent: CJ Entus.

The play-by-play announcers (two male, one female) call up eSTRO's first player, Kim Won Ki.[3] He rises from the counter, clutching his own keyboard and mouse, and climbs a small set of stairs, entering the gaming pod. It resembles a flimsy amusement park ride with an angled front windshield and large Plexiglas windows on the side. At the same time, CJ Entus's player does the same, stepping into his own gaming pod.

Inside each pod are two PCs and a pair of bulky monitors that take up nearly half of the "desk." The pod is set up for one purpose and one purpose only: to play StarCraft. Kim arranges his keyboard and mouse, and his opponent does the same on the opposite side of the stage. Each player pops in his earbuds to listen to the game's sound effects, and then covers his ears with a set of industrial ear mufflers to mask the sounds of the crowd and play-by-play reports, despite the fact that the pod is nearly soundproof. While the two opponents settle in and prepare to destroy each other with nothing but mouse clicks and keystrokes, the zebra-striped referees stand patiently behind each competitor. A turquoise neon light glows underneath each pod.

The crowd is surprisingly polite and quiet for a nationally televised sport—there are no chants, nor rhythmic clapping in anticipation of the match that is about to begin. The trio of announcers sits onstage, between the two pods, facing the crowd. Projected on a large screen above the announcers is the action of the game itself. The opening screen allows the players to configure the game and choose their race (Terran, Zergling, or Protoss) and finally this round's map.

In most developed countries, console games (Nintendo Wii, Microsoft Xbox 360) have tended to dominate over classic computer games played on a PC. In the United States alone, the console game industry is worth over $1 billion.[4] This is not the case in Korea, mainly because Japanese-made consoles by Nintendo and Sega could not be sold in Korea.

During the Japanese occupation of the Korean peninsula from 1905 until 1945, use of the Korean language was highly restricted, and Japanese-language media dominated the country. The Japanese imperial government made a concerted effort to supplant the native Korean culture with its own through the use of linguistic and cultural repression, and much more direct and harsher moves, such as the forced recruitment of Korean women to become prostitutes for Japanese imperial forces. As a way to counteract the historical imbalance between Korean and Japanese media, South Korean passed several laws after gaining its independence that restricted Japanese media and consumer products. As a result, the console gaming boom born out of Japanese video game companies like Nintendo and Sega in the mid-1980s and early 1990s did not exist in Korea. One could obtain a black market Nintendo in Korea in the 1980s, but it wasn't available on a mass scale. Korea did not lift its ban on Japanese products until late 1998.[5] Consequently, the only video games that were available for a long time were PC games. As a result, in families that had computers—ostensibly for educational purposes only—younger Koreans became more and more familiar with using a PC and the games that came along with them.

STARCRAFT IS KNOWN AS A REAL-TIME strategy game, in which players must collect resources, build military units, and seek out their opponent in the game's online map. As a player explores the map and discovers more resources, he or she will eventually discover the opponent's base, and then it will become a race to see which side can overwhelm the other with his or her respective armies. At a decent amateur level, games typically last thirty to sixty minutes; here, an entire one-on-one professional match usually takes less than fifteen.

No one is sure exactly why StarCraft continues to be so popular in Korea, though there are a number of theories. Some StarCraft watchers, like

Daniel Lee, the then-coach of the eSTRO team, suggests that the game itself is near perfect, as it is imbued with a profound, complex logic, like chess.[6] He also points out that the game speaks to the Korean idea of *taegeuk*, also known as yin-yang, which is symbolized in the blue and red swirl on the South Korean flag. "Another factor [as to why] StarCraft is so popular is because of the three races, they are so well balanced, one race doesn't have a great advantage over the others," Lee says.

Other Korean scholars have compared the game to capitalism itself. "The principle of 'maximum profits from minimum capital and labor' is perfectly realized in StarCraft. Just as capitalism turns out to be the most suitable regime in history so far, real-time strategy games like StarCraft dominate the game market," writes Park Sang-u of Yonsei University in his 2005 Korean-language book *When Games Talk to You.*[7]

StarCraft has even gotten the attention of the country's military leadership. In 2006, one of the Pro League's top players, Lim Yo-hwan, better known as SlayerS_BoxeR, had to serve his required twenty-seven months in the military, as all young South Korean men do. Lim had been a professional gamer since 1999, starting at the age of nineteen. However, once he was drafted into the Korean Air Force, the top brass sponsored the creation of the Pro League's twelfth team as a way to recruit for their military division, and possibly to get some young computer-savvy military minds in their ranks. Traditionally, the country's large media, telecommunications, and shipbuilding *chaebols*—Korea's government-directed conglomerate mega-corporations, including SK Telecom and Samsung—sponsor the other Pro League teams.[8]

CJ ENTUS AND ESTRO sit between the pods and the audience, at their respective clear plastic-topped counters. Under the counter, computer screens project the current match, allowing the other teammates to monitor the evolution of the game, studying it intently for possible tactics and strategies.

As lights dim and the music fades, a lone girl's staccato voice pierces the din in accented English: "One! Two! Three!"

"Entus Fighting!" responds the cacophony of mostly high-pitched female voices.

Not more than a second later, another voice pipes up from the other side of the audience: "One! Two! Three!"

"eSTRO Fighting!" bellows the response.

And with that, the StarCraft match begins. Game on, Korea.

Korea's Technological Landscape (2009–Present)

Boasting the first, largest, and most widespread professional "eSports" league, with supporting infrastructure such as corporate sponsors, the eSports Stadium, and its associated television networks, Korea has developed an Internet culture unlike any other, and the popularity of eSports is just the beginning. Many young Koreans are initially exposed to games like StarCraft by spending their free time in "PC Bahngs" (PC Rooms), which usually cost $1 US or less per hour and abound in every city.[9] PC Bahngs resemble cybercafés in many other countries, as they typically offer a cheap, high-speed Internet connection. But here, instead of sending e-mail or watching YouTube, most kids play online games either by themselves or in a more social atmosphere. Kids—boys, mostly—sneak out from under the watchful eye of their parents to put in an hour or three blasting away at virtual enemies across the room, across the country, or across the globe.

PC Bahngs provide a common form of escapism for kids who experience enormous academic pressure, both in their regular schools and in *hogwans*, the after-school subject academies. StarCraft gained currency and popularity with Korean youth because of the availability and affordable pricing of PC Bahngs. One homegrown game, NCsoft's Lineage, has also gained widespread interest, becoming one of the most popular subscription-based games in the world, with more than 400,000 paying monthly accounts worldwide, including 250,000 in Korea.[10]

In Seoul, like in many other Korean cities, PC Bahngs line the streets, advertising from large vertical electric signs that hang from towering buildings so they can be seen a block away. These skyscrapers and office buildings overwhelm the cities and blaze with light all day and all night—being in urban Korea can feel like walking into a science fiction novel. Seoul is a constantly moving city dripping with neon lights, the world's fastest Internet access, and state-of-the-art mobile phones in everyone's pocket. It is a place where anyone can watch a live baseball game on a cell phone while riding the Seoul subway at high speed hundreds of feet underground. By comparison, it wasn't until December 2009 that riders on the San Francisco Bay Area subway network, locally known as BART, could send text messages and make phone calls in the underground TransBay Tube between San Francisco and Oakland.

But Korea hasn't always been like this. Less than sixty years ago, the Korean War devastated the country. By the end of the war in 1953, three-fifths of the country's cultivated land was destroyed and property damage was estimated at $3 billion US. Ten million people, or one-quarter of the population, were left as homeless refugees in the wake of a peninsula-wide war. The final death toll exceeded 3 million. In the aftermath of the war, the Neutral Nations Inspection Committee wrote: "The country is dead . . . there is no activity. . . . The cities are completely destroyed."[11]

Yet, by the end of the century Korea had undergone the fastest economic development that the world has ever seen. Between 1963 and 1974, the country's real gross national product more than tripled, growing at 10 percent per year.

In that same year, per capita GNP hit an estimated $470 US, double what it had been a decade previous. By 2008, Korea's per capita GNP had surpassed $20,000 US.[12]

During the 1970s and 1980s, one important element of this rapid industrialization process was the strong top-down leadership to implement a strategy of long-term, export-oriented electronics. By 1988, Korea sold $23 billion worth of consumer, component, and industrial electronics, making it the sixth largest producer in the world. Today, Korean companies are among the leading manufacturers of televisions, monitors, and mobile phones. During the 1980s, Korea increased its teledensity (a measure of the number of telephones per one hundred people) faster than any other country in the world. In 1999, Korea became the first nation where mobile phones outnumbered landlines. The United States did not achieve this until 2005.[13]

Similarly, between 1995 and 2000, the number of Internet users in Korea grew fifty-fold. By December 2000, Korea had already reached over 19 million Internet users out of a population of 50 million—or two out of every five. According to the Organization for Economic Co-Operation and Development (OECD), Korea ranks in the top five of broadband subscribers per one hundred inhabitants, and is the only non-European nation besides Canada in the top ten. A 2008 OECD broadband report stated that Korea had "achieved the vast majority of its penetration level before 2003," far earlier than many of its peer countries. The report noted that "a combination of policy, geography, and competition put Korea far ahead."[14]

In short, Korea has transformed itself from a "dead" country in 1953 to become the most wired country on the planet a half century later.

While the adoption rates of hardware are faster in Korea than in the United States, Korea has pointed the way toward the future in online software applications as well. Today, in the United States, Facebook and MySpace are among the most popular social networking websites in the world. Facebook alone boasts an "active user" base of more than 500 million people. However, well before the founding of Facebook (2004), MySpace (2003), and even Friendster (2002), Korea already had an extremely popular social networking website, called Cyworld, which was created back in 1999.

Cyworld can be thought of either as a more elaborate version of MySpace, or a slimmed-down version of the virtual three-dimensional online world Second Life, which can even be used on mobile phones. Each member signs up for a mini-*hompi* (mini-homepage), which can be customized using a three-dimensional character, or avatar, that lives in a virtual dorm-style room. Users then spend real money to buy virtual currency, called *dotori* (acorns). One acorn is worth about $0.12 US—and six acorns will buy you a decent-looking couch. Cyworld users can also easily upload music, photos, or videos. Cyworld grew so popular in Korea that by 2006 it had 20 million users, or approximately

40 percent of the country's population, and an estimated 96 percent of the population under thirty.[15]

In the United States, social networking is also viewed to have had a profound impact on American politics—Barack Obama is viewed as the first "Internet President." After all, the Obama campaign created and successfully leveraged online social networking tools and even an iPhone application.[16] However, many Americans either forget or are unaware of Roh Moo-hyun, the first president or prime minister anywhere in the world whose election was directly influenced by the Internet.

Two years before the "skinny kid with a funny name" took to the national scene in 2004, underdog Roh Moo-hyun was the challenger in Korea's December 2002 presidential election. Roh's opponent was GNP candidate Lee Hoi-chang. Roh, who at the time was the minister of maritime affairs and fisheries, was a member of the liberal Millennium Democratic Party.

However, 2002 was the first Korean presidential election during which a significant portion of the Korean population had access to broadband service, and many people spent their time reading a two-year-old Korean news site, OhmyNews.com. The site covered Roh Moo-hyun's campaign unlike any other traditional media outlet—broadcasting and publishing his speeches and campaign rallies in their entirety.

OhmyNews provided a fresh perspective, as older, more politically conservative newspapers, including the nation's largest national daily, *Chosun Ilbo*, dominate the Korean media landscape. OhmyNews appealed to the younger, wired, and more politically progressive generation. By contrast, Korea's older and more conservative population typically supports the historically strong relationship with the United States, which has a significant number of troops stationed in their country, constantly ready to defend it against a North Korean attack.

During the 2002 Korean presidential election, OhmyNews was able to steer much of the political discussion to issues that were more favorable to the liberal candidate, Roh. These included alleged draft-dodging by his opponent's son and the accidental deaths of two Korean schoolgirls caused by American military personnel. After the American soldiers were court-martialed and acquitted on charges of negligent homicide, many liberal politicians, including Roh, questioned the strong military alliance between the United States and South Korea. There were public calls to begin renegotiating the American Status of Forces Agreement (SOFA) in Korea—a position that would have been unthinkable in the past.[17]

Many Korean scholars have theorized that OhmyNews's coverage helped push the Roh campaign over the edge—he won the December 19, 2002, by a mere 500,000 votes out of over 33 million votes cast. Regardless of exactly how much of his support can be attributed to OhmyNews, President-elect Roh seemed to think that the site made the difference, as he gave his first domestic post-victory interview to OhmyNews—likely the first (and probably only) time that a major national leader gave his or her initial interview to an online-only publication.[18]

TODAY, THE DOMINANT APPLICATIONS of the Korean Internet are gaming, mobile access, social networking, and new media. However, the Internet did not change Korea as much as Korea changed the Internet.

Professional eSports and new types of media like OhmyNews are just two applications made possible by Korea's widespread and inexpensive broadband Internet access. Many times over the last several years, Korea has been atop the world in terms of broadband subscribers.

Moreover, Korea's particular uses and applications of the Internet directly resulted from the intersection of the nation's turbulent twentieth-century history—the Japanese colonial period, the Korean War, and the postwar half-century of rapid industrialization. When the Internet arrived in Korea in the 1980s, it was during a period of increased industrialization and a government-led effort in the export of computer hardware. A decade later, Korea launched one of the world's first broadband Internet service providers, which coincided with the country's worst economic crisis since the Korean War. Low-cost high-speed access, combined with a new computer game, StarCraft, pushed an entire new industry of PC Bahngs and gaming to a completely different level. In parallel, the ubiquity of high-speed access has pushed other applications, such as social networking and media, to the point that they can influence national politics. In fact, a major online political group was founded in a PC Bahng—apparently the twenty-first-century Korean equivalent of an eighteenth-century Parisian café.

Contemporary Korea is the setting for an entirely new and unique set of media, politics, entertainment, and even bold urban design, which have sprung from extremely fast, widespread, and cheap Internet access. This deployment of Internet technology is a small but significant slice of the larger story of Korea's information revolution, which has many players both in and out of the government. One of these was Chon Kilnam, who led and participated in many of these changes.

Chon Kilnam and Modern Korea Grow Up, Apart (1943–1979)

Now a sixty-seven-year-old semi-retired professor at Korea's preeminent technological and technical university, the Korea Advanced Institute of Science and Technology (KAIST), Chon Kilnam has been an active leader in the technological changes that have become intertwined with the nation's recent history. The research he and his graduate students conducted in the 1980s and 1990s has had an inextricable and profound effect on the technological landscape of the country. They were among a core team that helped to accelerate Korea from being a technologically simple nation to becoming the envy of the world.

Chon's character is a premier illustration of what can happen when a nationalistic drive combines with high levels of education to envision what the future can hold and to set out to create that future. Chon, from his early days as a government researcher, saw the potential transformative power that the Internet could have in Korea. His own technological ambition greatly mirrors the country's unparalleled transformation and development. After the Korean War, just as President Park Chung-hee used controversial, authoritarian methods to ensure the nation's development, Chon—though not nearly as severe— exhibited the same drive that would not accept failure as an option.

Professor Chon's office sits in the computer science building at KAIST, in Daejeon, about 100 miles south of Seoul. The building of professors' offices remains quiet, with a few graduate students walking around. Entering Chon's office, there's no way to know that this room hosts an important pioneer of the Korean Internet. Next to the door, a rectangular table dominates the center of the room, adjacent to a few bookshelves. Few pictures adorn the wall, nor is there memorabilia of years past. Anchoring the room is his computer, a Windows desktop, which overlooks the garden and the rest of the university below.

Like those of many academics, Chon's bookshelves are laden with old dissertations and records. When he pulls out a bound document to show how one of his students wrote his master's thesis in the 1980s on the first Korean-language e-mail program, he chuckles and smiles. It's clear as the dust flutters off of its pages that he doesn't dwell on the past very much. "Here," he gestures. "You can take this. I have another copy."

Chon speaks in an almost grandfatherly tone as he uses short, modest, but almost punchy sentences. Yet, despite the fact that he bore witness to vast transformation and innovation, his greatest pleasure remains spending time hiking and rock climbing. When he speaks, he leaves more unsaid than explicitly articulated—and sometimes the pauses fill with hearty laughter. This personality, of having a serious, forward-moving mind with a tinge of lightheartedness, may reflect his background. After all, Chon Kilnam was born in one of the most painful parts of Korean history: the Japanese occupation.

In 1910, Japan annexed and brutally occupied the entire Korean peninsula. Prior to this takeover, Korea had never been controlled by a foreign power. To ensure loyalty, Japanese authorities set out to impose their own culture on this newly subjugated territory. Korean students were forced to learn the Japanese language, twelve Korean political organizations were abolished, and Korean-language, pro-independence newspapers were shuttered. Meanwhile, the Japanese Government-General created over 1,600 special police stations across the peninsula and deployed two army divisions to enforce these new measures.[19]

By the late 1930s, as Japan extended control into neighboring Manchuria and southern China, it imposed the "League for the Mobilization of the People's Spirit," a government effort to force occupied Koreans to support the Japanese war effort. Koreans were forced to visit local Buddhist Shinto shrines and recite a loyalty oath to the Japanese emperor, and further, were required to take on Japanese names. Between 1938 and 1942 all Korean schools had to use the Japanese language.[20]

The early 1940s were a turbulent time, to say the least. Nazi Germany reached the zenith of its power, having conquered significant portions of continental Europe. Meanwhile, much of the planet was engulfed in fighting the Nazis, and their impact on colonial holdings in Africa, the Middle East, and East Asia. Following the attack on Pearl Harbor on December 7, 1941, the United States formally entered the war. Japan's objectives during the war were to take control of the natural resources of its neighboring and peripheral nations along the Pacific, and to establish, by force if necessary, a "Greater East Asia Co-Prosperity Sphere." Japanese forces quickly captured the Philippines, Thailand, Malaya, Singapore, the Dutch East Indies, and other island nations. The attack against the American fleet at Pearl Harbor was meant to neutralize the only military force that could counter this plan. However, six months later, the Battle of Midway marked the decisive defeat of the Japanese forces by the American navy. In April 1942, the Doolittle Raid against Tokyo and other Japanese cities bolstered the morale of the American military. Meanwhile, back in Washington, D.C., in January 1943, the Pentagon was dedicated as the new prominent symbol of American military power.

It was in this environment that Chon Kilnam was born on March 1, 1943, in Hyogo Prefecture in southwestern Japan—one of six children. His parents were from a rural region outside of Taegu, in the southern part of Japanese-occupied Korea. Like hundreds thousands of other Koreans, the Chon family sought jobs in Japan during the chaos of World War II, as Korea was a poor and largely rural country, while Japan needed more people to work in factories and shops as their young men had largely been sent to war. During that era, Japan represented a chance at financial success for many Koreans, and in many ways it was easier to get to than other, larger cities (like Seoul) in the northern part of the country.

When the Chon family arrived in Japan, they didn't want to work in a factory, as many others of their generation did, and quickly set up a wholesale and retail clothing shop near their house in the Osaka area. They tried to integrate themselves and their children into Japanese life as fast as possible. The Chons continued to cook Korean food at home, but to be "publicly Korean" at that time was asking for trouble with the local authorities—the threat of imprisonment loomed large. As a child, Chon Kilnam took all of his classes in Japanese, had only Japanese friends, and his parents only spoke to him in Japanese. In fact, for the first few years of his life, before the conclusion of World War II, he and his five siblings and parents were required to take Japanese names. And yet, Chon was still considered Korean, one of 2 million living in Japan. In fact, when Chon was three years old, authorities in nearby Osaka ordered all Koreans to register and carry an identification card with a photograph and fingerprints.[21]

In the aftermath of World War II, the Soviet Union moved into the Korean peninsula, north of the 38th parallel, to accept the surrender of occupying Japan, while American forces did the same in the southern part of the peninsula. This agreement had been hammered out by the Allied leaders at the Potsdam conference of July and August 1945, without the express consent of any Koreans. Each side began to build, supply, and arm both North and South Korea to further exert their influence over their respective territories. The Soviet government refused to let North Korea participate in a 1948 democratic election to determine the future direction of the entire peninsula, and consequently after the American forces pulled out from the south in 1950, North Korea, backed by the Soviet Union, and later China, invaded the south. Three years later, after a U.S.-led United Nations force expelled the Korean People's Army back across the 38th parallel, the Korean War ended nearly where it began, in a stalemate, and yet remained at the front lines of the Cold War.

As a child, Chon Kilnam excelled in math and science, and enjoyed solving complicated math and physics problems, largely because of their precise nature. He never got too excited about literature or other liberal arts—it was too undefined, too fuzzy. However, in the spring of 1960, when Chon was a high school senior, political events began to heat up in his family's homeland and he turned his attention to politics for the first time. On April 19, 1960, thousands of Korean students protested the deliberate killing of a student who had died the month before during anti-corruption protests.[22]

Many students and other younger generation Koreans were upset at the fact that earlier that year President Syngman Rhee, an American-backed staunch anti-communist, had won his fourth term in office in an election where he claimed that he had received 90 percent of the vote. Further, Rhee was widely believed to have orchestrated the landslide election of Gibung Lee to the supposedly independent post of vice president. On April 19, the students marched from Korea University to the Blue House (the presidential residence) demanding that President Rhee resign. Their numbers overwhelmed the presidential security detail, and 125 of the students were shot dead before the guards relinquished their position. The president resigned a week later and was taken out of the country on April 28 to Hawaii via an American CIA plane.[23]

As a seventeen-year-old, Chon would run home, toss his book bag in a corner, and turn on the radio to learn about news from the homeland that he never knew. He spent hours transfixed to the Japanese-language radio and television, listening to the turmoil happening just across the Sea of Japan.

While Chon didn't immediately understand the politics or politicians involved, something deep within himself compelled him to follow each step of the political turmoil, and the continuing anti-government protests.

In May 1960, different protests began in Japan as well. Many socialist organizers began protesting the Treaty of Mutual Cooperation and Security between the United States and Japan, which had been recently signed by the Japanese and American governments, but hadn't yet been ratified by the Japanese House of Representatives.[24] Thousands of left-wing protestors turned out, often violently, to express their disgust at their government's apparent acquiescence to a foreign power. Many of the demonstrators were university students in the Osaka area, and in some cases high school students as well. On one particular occasion, there were plans for a huge protest in downtown Osaka, with several thousand high school students protesting along with university students. Chon's high school was planning to send one of the largest contingents— more than five hundred students. Chon was the president of the high school association and was asked to speak at this rally, but he realized that he could not speak for what the Japanese should do, as he himself was Korean. In other words, this wasn't his fight. He was more concerned with what was going on back in Korea.

"It was a big thing to the Japanese, but to me, I don't belong here," Chon recalled.[25] "I thought I was in the wrong place." He concluded that Korea was where he belonged. Although he was born and raised in Japan, and spoke Japanese as well as any ethnic Japanese, he knew that he would always be a *Zainichi Chōsenjin*, a Korean in Japan. He felt that the "real" Japanese people would always treat him differently, as a foreigner, and he felt that there was no place for him to advance in the society. "Japan is a fairly closed country," Chon added. "Foreigners have to be foreigners, and if you are born there, then being a foreigner all of your life is not too easy."

Quickly and decisively, Chon determined that once he had completed his studies in Japan, he was duty-bound to help Korea in any way possible. "I should contribute to the buildup of Korea," he said. "That's the reason why, initially, I thought about being a medical doctor, because a medical doctor is always needed and that's an easy way to help. But I didn't like those smells [of hospitals and medical equipment]. And I like mathematics, so engineering is a fairly obvious choice, because mathematics doesn't help much any developing country, the application is different. So I chose information technology, which is the closest to mathematics."

Chon felt that his parents didn't take him seriously at that age—they hoped he might put his math skills toward a degree in economics or business so he could eventually return to his home city and help them further expand the store. For them, Korea was a place that they had long left behind. It was a poor, inferior country that, as much as they loved it, offered them nothing to gain.

While a young Chon started to develop the seeds of his Korean nationalism and knowledge of computer science, his homeland was on the fast track to industrialization, led by President General Park Chung-hee. The president previously took power by force during a military coup d'état on May 16, 1961. At the time, the southern half of the peninsula was largely agrarian and rural. North Korea, backed by the Soviet Union and its heavy military-industrial complex, was considered the superior power. But President Park put the country on an irreversible course with his five-year plan. In 1962, when per capita gross national product (GNP) was $87 US annually, the president launched an aggressive five-year industrialization and export-oriented plan. The low level of industrial activity at that time can be seen in the number of registered cars: 30,800 in the entire country. Thirty-five years later, there would be over 10 million.[26]

Under Park's leadership, corrupt high-level businessmen who controlled significant industrial capital and assets were rounded up and presented with an offer few could refuse; many of the companies were converted to the nationalistic cause of building for the good of the nation. These *chaebols*, as they came to be known, received major benefits from the government, including access to subsidies and protection from union organization, in exchange for their agreement to follow the government's coordinated national plan. Companies that didn't agree to these terms or that could not meet their government-determined industrial production targets were threatened with government takeover. The Park administration also made it significantly easier for Korean businesses to have access to foreign capital through the Act for Foreign Loans (1962), which assured foreign investors that in the case of a loan default the amount lent would be covered by the government. Companies, meanwhile, put up their own industrial facilities as collateral to guarantee payment. By all accounts, the five-year plan was successful and laid the groundwork for Korea's heavy-industry "powerhouse period" of the 1970s and 1980s. During Park's administration, Korea became known for steel production and shipbuilding, and by 1969 per capita gross national income had more than doubled to $210 US.[27]

CHON TOOK HIS FIRST TRIP to Korea in August 1961 with his mother, who had not been back to her home country in decades. Although flying from Osaka to Seoul takes only a couple of hours, for a young Chon, it was like stepping back in time. While Japan had peacefully prospered and was rapidly expanding its industrial base, Korea remained in a disastrous state. Since the end of the Korean War, ordinary citizens had hardly seen a material change in their standard of living. Even his grandparents' house in Keochang (Kyungnam Province) was a crumbling hovel.

"It was a very heavy military environment," Chon recalled. "From the airport, you could see so many of those soldiers all over. So this [country] was a combination of student revolution and military—in the countryside, it was very poor. Around 1960, Korean GNP, [the] living standard was more like a typical country in Africa."[28]

At that time, Korea was a country trying to come to grips with its own identity, trying to forge ahead, striking a different destiny than one that would have been laid out by Japan. Photographs of Seoul from that time show only a few skyscrapers. Paved roads existed, but only in the city centers, certainly not in the periphery or in the rural areas. Seeing this land that was decades behind Japan made Chon question his own judgment.

"It was kind of new to me," he remembered. "In many ways, it gave a very strong impression for me. It was almost overwhelming. It was so different, the place that I decide to come to after education. Am I serious? Can I really survive here?"[29]

But Chon remained determined to prepare himself adequately and began his studies in engineering at Osaka University that fall. Four years later, when the time came for Chon to decide where to pursue his doctoral degree, his parents weren't much help. They gave him vague suggestions that he should get out of Japan and should attend school in the United States or Europe to broaden his horizons. He briefly entertained the idea of going to Moscow State University to study mathematics, but as his apt classmates pointed out, spending time in the Soviet Union would likely make things difficult for someone who wanted to later live in South Korea. In the 1960s, at the height of the Cold War, spending time in the heart of the Soviet Union and then returning to avowedly anti-communist South Korea would likely arouse suspicion. At the time, computer science in the United States was still a very young field—Stanford University and University of California, Berkeley, both founded their programs in the middle of that same decade.[30]

From a distance, Chon evaluated doctoral programs in electrical engineering and information technology. His final choices were Columbia University in New York, Carnegie Mellon University in Pittsburgh, and the University of California, Los Angeles (UCLA). One of Chon's friends had told him that Columbia University, being close to Harlem, was "dangerous," and that he wouldn't like Pittsburgh, as it was a cold, steel mill town.[31] He knew that Los Angeles was warm year-round. So, based on geography alone, Chon chose UCLA.

As a young foreign student interested in computer science, Chon's accidental choice couldn't have been more fortunate. Only three years before his arrival, a newly minted UCLA professor named Leonard Kleinrock began working with the U.S. Department of Defense-sponsored Advanced Research Projects Agency (ARPA). Kleinrock was interested in something called "packet switching," a fundamentally new way to send data across computer networks in small chunks or "packets." The switching part meant that points on the network ("nodes") could transfer the packets along any possible route, so long as the packets reached their destination and were reconstituted in the proper order at the end.

Before Kleinrock's theories, it was generally accepted that data moving along a network required a direct, constant line between each point, or node. For example, when two people call each other on the telephone, they are directly connected. There is only one such path to connect at any given time, and if that connection is broken, the line goes dead. Plus, when the line is silent, that connection cannot be used by anyone else. When computers communicated with one another, to send data, instead of a human voice, they had to directly connect themselves to one another. Kleinrock suggested that if they could break up the data into smaller pieces, it would be easier to send and there would be no single point of failure—the data would still be transmitted even if some node along the way were suddenly taken offline. If there were a failure along the way, the final node could re-request packets.

However, at that time packet switching was an unproven technology, as no one had actually built a significantly sized network using Kleinrock's principles.[32] ARPA, meanwhile, was interested in Kleinrock's work in order to build what became the ARPANET, a network of computers scattered throughout the country run by government-funded computer scientists. Eventually, the ARPANET became the basis for the global computer network that we now call the Internet.

CHON FINISHED HIS MASTER'S degree in computer science in just over a year, in December 1967—although the formal organization of UCLA's computer science department would not come for another year. After finishing the degree, Chon decided that he needed a break from academic life, so he took a job for two years as a system engineer with Collins Radio (now Rockwell International), whose business largely consisted of building high-quality radio equipment for the military and for Earth-to-space uses. His job was to advance a project similar to Kleinrock's ARPANET—Collins was working on a way to build a distributed computing system that could communicate wirelessly between land, air, and space units, based on packet switching technology. In essence, it would have been a private, commercial Internet that could work not only between computers connected via a physical data cable, but also to planes and spacecraft that were moving at high speed. Had the project been successful, it would have been sold to large companies, and probably the American military as well. However, computer processors at that time were not powerful enough to support a network of that scale and scope. Chon says that the project was ahead of its time by two decades.

"At that time, we just barely had the idea that packet switching may work. Or it may not work," he recalled. "It was amazing how they ended up having such an ambitious project."

This packet switching network project was his first job, and proved crucial for his professional development. Chon was exposed to ambitious thinking, and took on an undervalued project. His colleagues at Collins were convinced that computers would grow, that there would be more than just a small handful in each university or university department. As computers were the size of people at that time and cost thousands of dollars apiece, the notion of a personal computer was ludicrous.

In the year before Chon returned to UCLA, some graduate students and faculty at the initial four ARPANET institutions (UCLA, Stanford Research Institute, University of California, Santa Barbara, and the University of Utah) formed part of a research group interested in understanding the nature and architecture of computer networks, the Network Measurement Center (NMC). Another one of Kleinrock's graduate students, Vint Cerf, was also a senior member of the NMC.[33]

Chon decided to return to UCLA in the fall of 1969 to finally complete his doctoral degree. Many members of the NMC, including Kleinrock, Cerf, and others, were in the final stages of establishing the first ARPANET connection between two computers hundreds of miles apart. Again, Chon's timing could not have been better, as on October 29, 1969, the first message—"lo"—was sent between the two computers on the ARPANET.[34] By the end of the year, two more machines, or nodes, were added to the nascent network, one at the University of Utah and another at the University of California, Santa Barbara. These four computers were the first to be connected on the network that would evolve into the Internet.

As a foreign graduate student, Chon could not directly participate in the ARPANET, since it was 80 percent funded by the U.S. Department of Defense, and required American security clearance. And with the United States then at the peak of its involvement in Vietnam, Chon felt uneasy about participating in research funded by the American military. "As a foreigner, I didn't want to get involved in other countries' military," he said.[35]

This was not a trivial decision. At the time, the bulk of all engineering projects received funding from the American Department of Defense, and Chon had to find an engineering project that he could work on that would keep him separate from the rest of the department.

Chon completed his doctorate in 1974, five years before his wife, whom he met at UCLA, finished her Ph.D. He taught for the next two years at local universities and colleges, and two years after that, in 1976, he took a job with NASA's Jet Propulsion Laboratory in nearby Pasadena. He worked as a systems analyst at the Mission Control Center, and upgraded its computer systems, including the next generation of networking and user interface. His work at JPL furthered his interest in networking technology and exposed Chon to different types of computer hardware, but he mainly continued research in more practical engineering tasks. Within two years, he published scientific papers with titles like: "On Optimal Regulation Policies for Certain Multi-Reservoir Systems," "Performance of Automated Mixed Traffic Vehicles," and "Information Processing in Electricity Distribution Systems."

WHILE CHON WAS IN CALIFORNIA, Korea continued along its path of rapid export-oriented industrialization under the leadership of President Park Chung-hee. During the 1960s, however, Korea became a major exporter of radios, televisions, and later calculators and microwave ovens. The Electronics Industry Promotion Law of 1969 encouraged the sector with low-interest loans, and dropped production machinery duties to zero. As an example of the success of this program, Korean manufacture of televisions went from 10,000 per year in 1966 to more than 1 million in 1975, and while only 2 percent of Korean televisions were exported in 1968, by 1975 half of the production run was sold abroad. On a larger scale, total production of consumer electronics was worth $10 million in 1965, and had grown to $462 million by 1973, with an average growth rate of 60 percent per year. In addition, 80 percent of manufactured consumer electronics goods were exported. However, while Korea surged economically, Park had effectively made himself dictator of the country through a series of semi-legal reforms.[36]

The Park administration established many science and technology research entities, and created legislation to focus the country toward the cutting edge of consumer electronics. These included the Korean Institute of Science and Technology (KIST, 1966), the Ministry of Science and Technology (1967), the Electronics Industry Promotion Law (1969), the Korean Advanced Institute of Science (1971), the Law for the Promotion of Technology Development (1973), the Korea Institute of Electronics Technology (KIET, 1976), and the Korea Telecommunications Research Institute (1977), which later became the Electronic Telecommunications Research Institute (ETRI).[37]

One of the most productive of these research facilities, KIET, was largely created to prove that semiconductors (and eventually, computers) could be manufactured in Korea. Until that time there had been small-scale technology transfer between Korean expatriates in the United States, and offshore manufacturing by American companies, but nothing that had been entirely engineered in Korea.[38] Further, as Japan geared up for consumer electronics development in the 1970s and 1980s, Korea was eager to show that it could compete on a global scale with its longtime rival and former colonial power. The 1970s also marked the birth of the consumer electronics age for American companies, as Intel released its first commercial microprocessor, the Intel 4004, in 1971.

In addition to providing direct government spending for research, the 1972 Technology Development Promotion Law required Korean tech companies to reserve up to 20 percent of pre-tax profits for research and development. That research money—which was tax-deductible—needed to be spent within two years on any activity related to development, importing of foreign technologies, training, or donations. By 1978, 153 firms reserved over 30 trillion won (around $42 billion US at the time) and used 80 percent of that by the end of the decade.[39]

As a way to further this strategy during the mid-1970s, the Korean government also recruited Koreans who had recently completed their doctoral degrees from universities in the United States. With American engineering jobs drying up at that time, it made sense for many to return home. In fact, the first eighteen researchers recruited to work at KIST had all acquired their doctorates in science and engineering and had five or more years of research experience in the United States. By 1986, the Korean government employed 53,000 researchers—up from 6,000 in 1973.[40]

Nationalist pride also motivated many young Koreans who wanted to help rebuild their home country. One of these was Dr. Kang Ki-Dong, a Korean-American engineer who received his doctorate at Ohio State University and returned to Korea to found Korea Semiconductor in 1974. The company began full operations the following year, and brought large-scale integration (LSI) technology from Kang's former American employer, Integrated Circuits International, Inc. LSI permits tens of thousands of transistors to be placed on a single chip, and is the basis of many electronic devices, ranging from cheap pocket calculators to early microprocessors. This marked the first time that Korean companies could produce semiconductors on their own, and it operated in parallel to the similar rise in Taiwan.[41]

Also in the 1970s, KIST hired Dr. Lee Yong-Teh—then a young scientist who had recently earned his doctorate in physics from the University of Utah—to create a national semiconductor and computer development lab. Lee was convinced that Korea could be a powerhouse in the manufacture and mass production of semiconductors and eventually entire computers, not mere components. He had spent countless hours in front of the computers of the 1960s at the University of Utah, one of the original four nodes on the ARPANET. Lee knew then that while the "minicomputer" was on the path to great success, there would be a similar push in the "microcomputer" sector as well.[42] Both of these machines eventually came to be known as the "personal computer"—one for every person. (While it's obvious today why everyone would want their own computer, when computers cost a small fortune and took up entire rooms, this notion of individual use was not at all evident.)

Lee figured that with more overseas-trained Korean engineers like himself and enough government funding, it wouldn't be long before Korean companies would be atop the world for semiconductor, processor, and eventually computer production. "I started to claim to the Korean government: 'If you can give me 100 engineers, in three years I can build the best computer in the world,'" Lee recalled. "They didn't believe me."[43]

Chon Kilnam met Lee, a friend of his father-in-law's, for the first time in 1974. Lee was then part of a KIST delegation that came to visit technology companies in Los Angeles. The men became fast friends, and agreed that as soon as Chon's wife finished her doctoral degree from UCLA in 1979, Chon and his wife would move to Seoul and Chon would join Lee. This was the ticket back to Korea that Chon had been waiting for since 1961, a way for him to return to his ancestral homeland and help it along the path to development.

The pair also agreed that Chon would stay for a few years, after which he would be allowed to leave and pursue a career in academia. While Lee waited for Chon, he helped to kick-start KIET, which was one of the components of a 1976 recommendation by the World Bank and became a keystone of the 1977–1982 (Fourth) Five-Year Plan. As this report states:

> For aggressive penetration of the consumer electronics market, the growth target and plans formulated by the government and industry must be supplemented as soon as possible by a comprehensive master plan for the development of the industry. This plan should integrate market research, product strategy, diffusion of technology, marketing strategy, and resource requirements.[44]

Lee did exactly that. He built KIET from the ground up, hired dozens of engineers, and managed to get the research institution from a pipe dream to a functioning semiconductor plant within a few short years. Lee later said that the level of production quality, at that time, was "the best in the world." Flush with $24 million in World Bank loans and government grants, KIET had plenty of money to acquire new equipment and consultants, and to rapidly build Korea's technical base. In 1978, KIET established a liaison office in Silicon Valley, which was crucial for acquiring capital equipment to export back to Korea, for obtaining technology licenses, and for staying abreast of what was happening in other parts of the industry. Toward the fall of 1979, Lee looked forward to having Chon help run KIET and take it to the next level.[45]

However, on October 26, 1979, shortly after Chon's arrival back in Korea, President Park Chung-hee, the architect of much of Korea's postwar development, was assassinated. The killer was the president's colleague and friend Kim Jaegyu, the head of the Korean Central Intelligence Agency, who believed his actions to be a protest against Park's consolidation of power. Kim was executed the following year as punishment for the assassination.

Chon Ignites the First Korean Network (1980–1983)

The death of Park marked the end of an era of staggering growth and the pinnacle of industrial activity. In a country where people had greater education and increased economic freedom, it was only a matter of time before they would want greater democratic freedom as well. In the years before, Park had virtually become a quasi-benevolent, albeit legal, dictator. In the immediate aftermath of Park's death, Prime Minister Choi Kyu-hah assumed the office of the president. Due to various political struggles from within the government and the military, he lasted barely a year before a military dictatorship took over Korea. General Chun Doo-hwan became president by September 1980.

Under the Chun administration, government spending was quickly reined in, large-scale industrial credit was tightened, and the government allowed more market liberalization.[46] This marked a distinct move away from the Park-style interventionist policies that were responsible for such a rapid industrial buildup in Korea. The new government downsized much of the existing bureaucracy, and merged similar entities, like ETRI and KIET. Lee Yong-Teh felt that this merger diluted the goal of KIET, by incorporating unrelated telecommunications research. So he quickly resigned, leaving Chon in charge of the systems division of KIET (the other divisions being semiconductor design and industrial processes).

That same year, in 1980, Lee quickly formed a new company, TriGem, which later became Korea's first commercial computer manufacturer. By 1982, the Korean government also asked Lee to head up a new government monopoly company to be in charge of Korean data communications, largely for the government itself—hence its name, Dacom (Data Communications). But Chon remained an omnipresent figure in the Korean tech sector. "Chon Kilnam was always my advisor," Lee recalled.[47] "He is the first one who was aware of the importance of the Internet."

Meanwhile, KIET's systems division was supposed to figure out how to build a computer, domestically. However, despite Korea's ambitious results and near-term goals, the design and manufacture of a computer was a massive undertaking. At that time, only developed countries in Europe, North America, Taiwan, and Japan had the technical knowledge, not to mention the financial resources, to construct a computer at $100,000 apiece. At the time, Korea's per capita annual income was under $2,500.[48]

But it was Chon's job to try to figure out how to make it work. When he arrived, there was already a rudimentary plan for a viable computer system that could easily be made and sold by Korean companies to buyers both domestic and abroad. After all, this export-oriented strategy was exactly what had made Korea's early consumer electronics, textile, steel, and shipbuilding industries work so well. At that time, Japan was building some of the fastest computers in the world, and the Korean government was willing to do whatever it took to both boost the domestic economy and outperform its former colonial overlord and rising neighbor.[49]

Two main factors informed Korea's nationalistic drive to overtake Japan. The first, of course, had to do with its half-century of occupation, which included restrictions on business. In 1911, Tokyo passed legislation requiring the Japanese Government-General approve all new companies and businesses in Korea, which had the effect of essentially halting all non-Japanese owned new business and economic activity. As the occupation period continued, Korean nationalism began to use business as a way to assert independence and self-reliance. The Korean Products Promotion Society was founded in Pyongyang in 1922 to encourage Koreans to buy Korean-made products. At the time, Korean nationalist elites believed strongly that the road to independence from Japan relied substantially on modernizing its own businesses, infrastructure, and technology.[50] However, those hopes ultimately were dashed as the colonial government enforced harsher and harsher laws against Koreans in the 1930s and 1940s.

The second factor, while related, may be explained by the political economic concept of "relative backwardness." This was a political theory espoused by the twentieth-century Russian-American economic historian Alexander Gerschenkron, which states that if a particular economic system lags behind a neighboring one, the disparity spurs politically motivated and institutionally guided innovation. Initially, Gerschenkron applied his theories to Europe and the Soviet Union, but they are also clearly applicable to South Korea. Being relatively backward compared to Japan created intense political pressure domestically, particularly when Japan was creating its semiconductor and, later, personal computer industries. Seoul's push for many semiconductor factories to compete with those from Japan worked initially, and with an institution like KIET, Korea had finally merged theory and practice in one physical location. Those factories could easily then be transferred to local *chaebol* corporations. In short, Gerschenkronian innovation policy combined with a desire to settle political-historical scores has yielded a rather productive country.[51]

In fact, Korea had a name for its newfound state policy of technology acceleration and production: *Gisul Ipgook,* or "nation building with technology advance." Several times a year, the government convened an "Expanded National Conference for Technology Promotion," as a way to highlight the collective goals of semiconductor and computer production to the government, investors, academics, and journalists.[52]

Chon Kilnam was one of a number of key figures in the *Gisul Ipgook.* As he took over the head of the computer systems division, he was tasked with making sure that Korea could produce what was then called a "microcomputer"—which became the personal computer. "In Korea, it was not obvious," Chon remembered. "[We wondered:] 'Is Korea ready for computer development?' Computer development only belonged to advanced countries—only three countries in Europe, Japan, and USA."

In other words, it was a point of pride, being on this technological frontier, pushing the limits of what was possible. In the fall of 1980, the ARPANET had grown to dozens of nodes, nearly all of which were still in the United States, except for one computer in London and another in Norway.[53] While his colleagues at KIET and in the government were concerned with pushing industrial and commercial capability by manufacturing and selling computers, Chon had

foreseen the power of networks because of his experience studying and working in California—he knew that networks would allow unprecedented levels of
communication, and that Korea needed to be part of this global community.

In September 1980, atop a high-rise office building in downtown Seoul,
Chon was summoned to the Korean Ministry of Industry to present KIET's
annual strategic plan and budget. This was a continuation of series of quarterly
meetings that he and the KIET team had been having with lower-level bureaucrats. The primary goal of the government was to build semiconductor-
manufacturing capacity, with manufacture of a full computer as the secondary
goal. Chon acknowledged that KIET could and should build a computer, which
the Ministry agreed with—that wasn't the issue. Originally he asked for around
fifty engineers, as he estimated it would take the team one year to build a computer. Chon only got about forty engineers, who were paid around $20,000
each. But he figured that it would still be possible to complete the project even
with reduced manpower.

The real issue, Chon tried to explain, was the next step, the problem of networking the computers. He tried to illustrate how a group of computers that
had the ability to share files and send electronic messages could allow computer scientists to collaborate in an entirely new way. Of course, this required
diverting a small number of resources—engineers, their salaries, and budgets—
toward networking. But as Chon made his case for networking, all he got back
was blank stares. Computer networks, outside of the ARPANET, remained in the
realm of arcane academia. There was no obvious commercial or industrial
value, and thus no reason for the government to provide funding. Even as a
tertiary objective, it was basically out of the question.

None of the members of the Ministry asked questions—they just wanted to
build the hardware, and they wanted it done fast. They didn't want to be both-
ered with any other projects that would potentially sidetrack the main focus.
The request to build the network was denied. While Chon understood their
position of not wanting to build it, as there was no natural way to commercial-
ize this research—he was still "a bit pissed off."[54]

Although Chon did not yet have the official blessing of the government, he
had already begun research while working as an adjunct professor at Seoul
National University (SNU), where he had been teaching since his "return" to
Korea. He taught a graduate-level course in computer networking, largely as an
academic exercise, but also as preparation for a larger, national project that he
hoped would come soon.[55]

At the time, the Chun administration saw semiconductor production as
the key initial step to broader consumer electronics domination. In 1981, it
published the "Basic Plan for Promotion of the Electronics Industry," a document outlining how to connect both "upstream" and "downstream" components of the entire manufacturing process, from the smaller components all the

way through to the consumer electronics that they would be part of. American academics who have studied this period in Korean economic history have noted the "conscious imitation of Japan's structure of semiconductor companies' being divisions of larger electronics companies, themselves part of giant conglomerates," and that "Japan's recent experience . . . is very much in evidence in Korean promotion plans of the 1980s. The watchword now was coordination rather than state domination."[56] Getting semiconductor manufacture and quality up to speed and competing on a world scale was the priority, and the next priority was getting those semiconductors into computer memory chips (DRAM). In other words, the Chun administration and the Korean *chaebols* wanted a very measured, deliberate, and ambitious approach. If building semiconductors was step one, building a computer was step four and networking them was step six.

By early 1981, Chon managed to convince the Ministry that networking was necessary as a small component of computer hardware and software research. His pet project was finally approved, and was included in the budget from then on. He received "less than 5 percent, probably 2 percent" of the annual million-dollar budget. The project was called the Software Development Network (SDN), and was one of the few computer networking projects in the world at the time besides the original ARPANET. Chon had a full-time staff of one engineer, his graduate students from SNU, and a few more graduate students from KIST who came to work on the project. "It was my hobby. Since I am the director they cannot complain if I had a pet project," he remembered.[57] Chon spent most of his time and energy worrying about the networking project. On September 30, 1981, he penned the following memo:

> Computer networks as a software development tool is needed to carry on the National Project on computer development. The Software Development Network (SDN) will be used for the following purposes;
>
> 1. Memo exchange
> 2. Program (source and object codes) exchange
> 3. Computer resource sharing
> 4. Database access
> 5. System testing
> 6. Computer system development
> 7. Learn working under network environment
>
> We may be able to use our own computers for the network. This network may be similar to CSNet, Computer Science Network for universities. It is developed by NSF, U. of Wisconsin, and BBN.[58]

THE MEMO ALSO PREDICTED that the "basic system"—with two nodes on the network—would be fully operational by June 1982, and the "full system [would be] running by December 1982." The main "trunk line" between the two nodes would be at 4800 bits per second, or bps, a phenomenal speed at the time. (By today's standards, this is a minuscule speed.)

The goal of a nationwide and international computer network was simple: to ease the flow of information between computers on a global scale. It would allow scientists in various countries to be able to correspond, to share data and articles with one another significantly faster than they had been able to before. Chon felt that by networking all computers, the limit of data that could potentially be stored, and therefore known, could eventually become essentially infinite.

At the time, Chon and his team had to build everything from scratch. While they could have based their design on the existing hardware from the ARPANET, because it was a Department of Defense project the hardware and software could not be exported. In addition, the Korean crew was using a different computer than what UCLA had previously used over a decade before. In other words, Korea had to reverse-engineer everything. As Chon wrote in an academic paper, later, in early 1983, while describing the 1982 national computer project: "This project gives Korea an opportunity to catch up with the developed countries in some areas of computer technology in five to ten years and to compete with them for development of 'user-friendly computers' later on."[59] By user-friendly, he was referring to machines as we know them today: with graphic interfaces, pull-down menus, icons, and the like.

Beyond the hardware of the computer, the first challenge was to build a network router—a device that can compute when and how to send data to various nodes along a network. Today, a router can easily be bought for around $30, but three decades ago that would have been unthinkable. In 1969, armed with government grants, U.S. defense contractor BBN had built the Interface Message Processor (IMP), a huge device larger than a person. This $30,000 machine served as a standard device to connect to different types of computers at each site, and had software that defined particular rules and conventions for how to transfer data between them. IMPs could not be exported, so Chon and his colleagues had to make one of their own. Further, custom software was needed on the computers and routers in order to make it all work together. Chon obtained some prototype routing software from a California tech company, but it was very buggy and didn't work very well.

In essence, Chon and his colleagues had to reverse engineer the entire network. They couldn't simply bring the ARPANET machines to Korea, due to U.S. export restrictions. Plus, the hardware had to be homegrown, so that Korea could, in fact, compete with other countries, which made things even more complicated. Worse still, Chon and the team were trying to build a computer network when only one person among them (Chon himself) had ever even been around such a network before. It would be as if they had been told to build a house—having only seen a passing glimpse of one before.

Over the next few months, one of Chon's students was finally able to "ping" the SNU computer. Ping is a common Unix tool that allows one computer to send small amounts of data to another computer to see if they are connected properly. If the remote machine is connected, then it will respond back with the amount of time that it took for that data to be sent, expressed in milliseconds (ms):

PING lemonde.fr (195.154.120.129): 56 data bytes

64 bytes from 195.154.120.129: icmp_seq=0 ttl=237 time=353.719 ms

64 bytes from 195.154.120.129: icmp_seq=1 ttl=237 time=355.020 ms

From there it was only a matter of time before various other basic networking applications could be figured out, including e-mail, file transfer, and remote login.

One day in May 1982, one of the graduate students called Chon over to have a look at a possible breakthrough. Normally, the graduate students worked on a bunch of computers in a single room, while the professor kept his own office down the hall. All the eyes in the room turned to look at him as he entered. Chon sat at the computer, his fingers flying effortlessly across the keyboard to type out the single command: rlogin. This simple word allowed him to "remote login," or take control of, the SNU computer and access its files at a distance. It worked, just as he had seen at the ARPANET many years before. Then, in a strange loop test, he logged in remotely back to the same KIET computer from the SNU computer that he already was remotely logged into at KIET. Chon remembers it as being "very impressive."[60]

"Anything, after so many years of those suffering and working hard, and finally the thing works—you feel so good," he remembered. He and his team had done it—they had created the first TCP/IP network in Korea. This breakthrough of a two-node network happened one month ahead of schedule. Chon quickly called Lee Yong-Teh and a few other colleagues to share the good news, and he then took the group of researchers out for dinner and drinks. As the night went on, Chon, Lee, and the others began to sense that they were on the verge of bringing Korea into a new era, although they didn't know just how it would turn out. At the time, they knew that it was an important achievement in terms of this computer research project, but it was still just that: an obscure networking experiment. Just a year shy of Chon's fortieth birthday, he had helped Korea take its first baby step onto the Internet.

BY SEPTEMBER 1982, with the computer project well under way, and the networking project requiring broader and more widespread attention, Chon decided that it was finally time to leave KIET and start the academic career that Lee Yong-Teh had promised him. So he took a job at KIST (later renamed the

Korea Advanced Institute of Science and Technology, or KAIST) as a professor in the computer science department. The networking project continued under his guidance, and the project built the Han-16, the first Korean-built microcomputer, which was completed in early 1983 at KIET.

Between 1983 and 1987, Korea tripled its level of spending on basic research, reaching a high of 1.9 trillion won (around $2.7 billion at the time). As a percentage of national research and development, basic research was 17 percent of the state budget in 1986—a figure higher than that of the United States and Japan during that same period.[61] By May 1984, the Korean government-founded company Dacom began the first Korean-language commercial e-mail service. Moreover, in 1985, Chon helped to organize the Pacific Computer Communications Symposium in Seoul, one of the world's first conferences on the Internet.

Chon's leadership at KAIST not only created the first computer network and microcomputers in Korea, but it also encouraged an entirely new way of thinking. Because Chon had lived abroad and had experienced the industriousness and high quality of modern Japanese life and ambition, creativity, and discipline of the American educational and business culture, he wanted Korea to combine the best qualities of the two. By the end of the 1980s, Chon oversaw students who designed the first long-distance Korean computer network, created the first Korean-language e-mail program, and managed to get this network onto the experimental Internet—the first non-American network to do so.

But none of this would have been possible without choosing the right students.

Assembling the Team: Chon's First Generation of Graduate Students (1983–1990)

Park Hyunje remembers most of his graduate school interviews with KAIST professors as being quite easy. But Chon Kilnam was different.[62] Park had aced the entrance exam for the computer science department, but still had to go through formal interviews with members of the faculty. Professor after professor gave him softball questions, asking him only to present his academic and professional credentials. With such a high score, Park guessed that he could have said nothing of substance and still would have been selected. He had a stellar background and had just finished his studies at Seoul National University—one of Korea's top schools—with a bachelor's degree in sociology. Park wanted to further study human communication, and with computers being increasingly used as a way to correspond, it seemed obvious that computer science and computer-based communication offered the best way to merge these interests.

Unlike the other professors, Chon took the interview seriously, albeit casually, and asked him different questions on a single theme: how to get better students at KAIST. He wanted to know how to get KAIST to the level of institutions like the University of California, Berkeley, or the Massachusetts Institute of Technology—how to raise the level of thinking, scholarship, and research to world-class caliber. Park didn't have much of an answer, but said that KAIST needed to work as hard as it could to be the best university possible.

In the weeks that followed, Park went from lab to lab, spending seven days among each group of graduate students and their professor. The groups were friendly and welcoming, and each invited him out for a round of drinks at the end of the week—with the exception of Chon's group. Park saw that they were a much more serious bunch, and focused their efforts on working, not socializing. This group was tight-knit because of the long hours that they put in together in the lab, not because they had spent many nights downing rounds of *soju*. He knew then that Chon's lab—also the only one with its own computer—was where he belonged.[63]

That fall, Park joined Chon's ten other graduate students, the largest such group in the entire computer science department. Partly because Chon was a new faculty member, and partly because he was highly respected, there was no cap on the number of students that he could oversee. Chon didn't coddle his students; he gave them directions, and if they dropped out, so be it. He didn't really care how many of his students completed their doctoral degrees. The mission was nothing more and nothing less than to get the network fully operational, to understand its quirks and characteristics, and eventually, to figure out how to get it connected to other nascent computer networks outside Korea. The ultimate goal was to build a national computer network, along the lines of what the ARPANET had done decades earlier.

Chon's students were the best of the best computer science graduate students in the country. The government, via the school, had awarded them all full scholarships, free on-campus room and board, a small stipend, and the highly coveted exemption from military service. He would remind them nearly every day that they were the most privileged group in the entire country, and in exchange for that generous opportunity they should work appropriately hard.[64] "Think and act like a professional," he told them on an almost daily basis.

On a practical level, Chon changed some of the everyday organizational skills of his students. He insisted that his students keep a written record of everything that they did, and "he always emphasized that verbal communication is just passing things by," recalls Lee Dongman, Chon's first graduate student to complete his doctorate.[65] "If you want to convince people, you have to write."

Unlike every other professor in KAIST's computer science department, and likely in the entire country, Chon made his students write a properly formatted one-page progress report on their research each week. These memos, submitted alongside a verbal presentation, were collected at the department's meetings.

If even a single page was not properly formatted, Chon would immediately throw it away.

Kim Yoon, who managed the lab during the 1984–1985 academic year, compared Chon to a "drill sergeant," noting that it was in these meetings that Chon pushed his students to improve their English, and to study and attend conferences overseas. Lee also recalled that while Chon was tough and demanding, he was also very personable. "He treated his students not as a normal hierarchy, but as [peers]," he says. "[The fact that he] treated us as colleagues, that was a big difference. We could just sit down together and spend many hours and he gave us coffee in his office. [That] was very different from the Korean standard."[66]

The work schedule was similar to that of a late-1990s Silicon Valley startup: wake at 9 or 10 A.M., stumble down to the lab from the dorms, work all day, take a quick dinner break, work until midnight or later—lather, rinse, repeat. Chon's students didn't have weekends free, either. On Saturday mornings he would organize hiking, trail running, rock climbing, and skiing trips to the mountains around Seoul, and while he never explicitly told his students that they were required to go, cultural tradition demanded that when a professor invites you to something, you had to go, whether you wanted to or not.

"At that time, that particular time, basically we were looking for something, which is impossible in Korea, or nearly impossible," Chon recalled. "Doing the Internet back in the 1980s was just about impossible in Korea, so if I had to make it possible, I had to do something very special. I had to train them to be ready. To me, it's supposed to be obvious. We're developing professionals. Graduate school is for training professionals. We didn't have a culture of graduate schools, like MIT, or Berkeley or Stanford. They have their own culture. We had to have our own equivalent culture, in a positive way so we can compete against MIT or Berkeley. That's the process. They should prepare to be a professional. And because it's a long-term game, they have to be healthy. If you have to work ten to fifteen hours a day for four or five years, then your physical condition has to be good."[67]

Chon thought that if his students could solve problems on-the-fly on a rock face, trying to figure out where to grip, hold, or steady their footing, then surely they should and would also have the mental faculty to solve big computer science questions. He viewed the exercise as a key step toward making the Internet work in Korea, which in turn was a key step in the country's development.

"I had to realize the Internet in Korea, which is awfully difficult," he said. "If I fail to build up Internet, then the whole exercise doesn't mean much."[68]

After climbing, he would invite the students over to his on-campus home, and Chon's wife would cook dinner for the entire group. These dinners often went well into the night—and Chon's dinner table and camaraderie served as a surrogate family for the young students, who were all away from home for the first time. In fact, most students spent more time with the Chon family than they did their own parents and siblings. Hur Jin Ho, a classmate of Park Hyunje's, summed up his graduate student weekends this way: "Climbing on Saturday, and dating on Sunday."[69] In fact, it was through these Saturday-night dinners that Hur was introduced to his future wife.

At the dinners, Hur was mostly motivated by tales of Chon's childhood, growing up outside Korea, and later realizing that he was not only of Korean background, but was Korean in his heart. Chon told the group over and over again that whether or not they knew it, they had an element of noblesse oblige.[70]

"The thing [Chon] emphasized is that we have to keep thinking about what we can contribute to the society, in general," Hur recalled. "As privileged persons, KAIST students can be considered very, very privileged, in many senses: with exemption from military service and funding from the government and a very, very good working environment. He emphasized that you are so much privileged so you have to think about what you can contribute back to the society. That was the biggest thing he wanted us to know."

But not all the students took this lesson to heart. In 1987, after four years of rock climbing trips, Park hated them just as much as he did the first year. He wasn't the type to spend his free time outdoors. All that suffocating dust, the salty sweat, the unforgiving heat, the dreadful cold, and the pack that dug into his torso—he had had enough. Meanwhile, Chon, more than twice the age of these twenty-somethings, out-hiked and out-climbed them at every turn. Park's idea of a good time is relaxing indoors, usually around a dinner table with his buddies, knocking back a few *sojus*. Chon made sure that Park and his other students spent some time outdoors hiking or swimming nearly every day, a practice that he continues to this day. "I don't want to have fresh air even on the weekends," Park says, laughing at the stark contrast between himself and his mentor.[71]

With each step on the trail, Park's mind began to wander off, inevitably drifting to the job that consumed most of his time: being the network manager of the SDN. It was a difficult but ultimately rewarding job. When he first got to Chon's group, he didn't know the first thing about networking. But everyone needed a job, and the SDN needed a manager, no matter how green, so Park spent many hours pouring over the UNIX operating system manual, trying to figure out how to make it work. By December 1984, Park was able to connect the SDN for the first time to the Internet via another American academic network known as CSNET. This breakthrough allowed Korea, or at least a handful of computer science students and faculty, to send and receive e-mails to and from anywhere in the world.[72]

Over time, it became Park's job to make sure that everything ran smoothly. If e-mails somehow got lost, or were misrouted—a process that today is well understood and is highly automated—it was considered Park's fault. Because Park was in charge of the network, he had to write annual funding requests to corporations and to the government. But as the science itself was new, it wasn't immediately obvious how this computer networking would turn into a commercially viable product. Every budgetary meeting had the same alchemic outcome. Park would make his impassioned plea for raw materials (money), and the government committees expected it to be turned into gold, when in fact it was just a refined version of the lead that went into it. Each year required a Sisyphean effort to get even the most basic amount of funding for the SDN.

But Park didn't come to KAIST to run a computer network that he barely knew how to run. He came to do research, and to earn his doctorate. So while he was trying to establish network connections, troubleshoot its problems, and secure funding, he wasn't doing any research for his dissertation. Usually students in their third year begin their dissertation. Park was in his fourth year and still hadn't started yet. And then there were those stupid rock climbing trips. It wasn't worth it, he thought. All the stress of running the network, not being able to do the research that he'd come to do—he was ready to drop out. So he went to see Chon in the spring of 1987.

For Park, the decision wasn't easy. He agonized over the shame that he would have to face. Chon had instilled in him and the other students a sense of duty to the lab and to the country. Park and his classmates were the best and the brightest, and the country had sacrificed for their benefit—exempting them from military service in a country technically still at war, after all. Worse still, Park's family would suffer dishonor if their son dropped out of this prestigious program.

Park sat down across Chon's desk and began outlining his case. The litany of complaints came out quickly, and he almost stumbled over his words: he was in over his head with the networking, he didn't feel like he was doing a good enough job, and he had more work and more responsibility than any of the other students. He acknowledged that he made mistakes, probably too many, and hadn't devoted enough time to his classwork and studies. Surely he couldn't be expected to continue. Park could feel his voice growing colder and more distant as his body tensed up. But before he could open his mouth again, Chon looked at him and spoke first.

"You are good," Chon said. "Now it's time for you to start [your] research."[73]

Park Hyunje completed his doctorate in February 1990, then stayed on another few months to help with the network. In March 1990, he helped set up a 56 kbps network via satellite between KAIST and the University of Hawaii, which became Korea's first permanent connection to the Internet.[74] Prior to this, Chon's SDN had to dial in to networks in the United States or the Netherlands to send or receive e-mails, or to access Usenet, an early online discussion board.

The Rise of a Broadband Nation (1994–2000)

By the 1990s, just as the dot-com boom was beginning to rumble in California, the first generation of Chon's doctoral students had graduated. Chon encouraged his students to be entrepreneurial—and not to necessarily take a purely academic path—to take risks, and to explore life outside of Korea. When Lee Dongman, Chon's first doctoral student to graduate, finished in 1987, Lee took this advice and promptly applied for a job with Hewlett-Packard in California.

"I was the first person who was directly hired by an American company as a KAIST graduate," he said. "Many Korean students didn't believe it."[75]

They remained incredulous because Lee was part of the first generation of students that had been educated in Korea and could now compete on a global stage—starkly different from a generation before. Their professors, like Chon, of course, had been educated abroad and recruited to return. Even though their professors were good, many students and faculty in Korean engineering programs at that time still felt like they were playing catch-up. Many were not ultimately sure that Korea was ready yet to compete with a better-trained American and European workforce.

Going to work abroad was one different path, to be sure. One of Lee's younger classmates, Chris Chung, took Chon's advice and founded a startup in 1989, called Human Computer, Inc., or HCI.[76] This company was the first Korean-language desktop publishing company, a pioneer in using computers to manipulate text in the Korean alphabet, known as *hangul*. This was only seven years after the founding of Adobe Systems in Silicon Valley, the company often credited with sparking the desktop publishing revolution in the United States.

When Hur Jin Ho completed his doctorate the following year, Chung invited him to join his company. Hur had arranged for post-doctoral study in the United Kingdom, but gave it up to continue working with Chung after receiving an attractive offer to be part of the management staff. He was put in charge of developing Korean fonts and making sure that they worked well with laser printers.[77] The network of Chon alumni was beginning to plant its seeds in the burgeoning Korean tech sector.

When Hur graduated from KAIST in 1990, there was one major indigenous computer manufacturer, TriGem. Founded in 1980 by Korean tech guru Lee Yong-Teh, TriGem became one of the earliest and largest tech companies in the country, and had significant influence over many other companies that followed later, some of which were government run. Its role was not unlike that of Silicon Valley giants IBM in the 1950s or Hewlett-Packard in the 1960s and 1970s—a company where Apple cofounders Steve Wozniak and Steve Jobs worked as teenagers and young adults.

After working at Human Computer for two years, Hur was snapped up by TriGem in 1992 and was put in charge of "network business." He managed a team of forty employees who used computer networks as a way to sell more computers in Korea—they helped design many of the earliest corporate local area networks (LANs), as well as printing and file servers. At that time, Hur was one of a small handful of people in Korea who honestly could say that they had a decade of networking experience.

After working for four years, Hur recalled the guiding words of his professor and mentor, Chon Kilnam—that he and his classmates were the foundation of the Korean IT industry. They were there to become the future of Korea. Over and over, Chon told them that they could not simply work for existing companies, or worse, simply become computer science professors. Their mission was to create, to innovate, to think differently about what had been done before. Hur knew that the time was right for him to strike out on his own.[78]

Chon's almost evangelical mission, combined with Hur's interest in America's tech sector—he had been reading up on the story of Apple cofounder and current CEO Steve Jobs—made him anxious to start something entirely new.

"I had some wishes to start my own businesses in the same manner as [Jobs] did in Silicon Valley," Hur said. "But I didn't have a clear idea of what that meant. There was no concept of a startup in the early nineties [in Korea]. But I wanted to do something myself."[79]

As the SDN developed under the leadership of Chon Kilnam in the 1980s, government-funded firms, including Korea Telecom and Dacom, began serving as the early Internet service providers (ISPs). However, given that such access was very expensive, and that these institutions were stymied in their own internal bureaucracy stodginess, Hur knew that Internet access would not spread very quickly. Hur believed that he could do a better job. He decided that with funding from TriGem, he would create the first private ISP, skirting existing Korean telecom law—pushing the limits of the possible.

At the time, Korea had two types of telecommunications licenses, one called a "carrier license" and another called a "value-added license." The main difference was that a carrier was defined as running on its own infrastructure, whereas a value-added license added service on top of other companies. Hur had been following the development of the Internet in neighboring Japan, and knew that IIJ—the first Japanese ISP—began its e-mail-only Internet service in July 1993, followed by a commercial Internet service in November 1993.[80] However, he knew that IIJ spent two years waiting for the Japanese government to figure out what type of license the company needed to operate as an ISP.

"I believed that the Korean government would do the same thing and kill [the project]," Hur remembered. "So I got the value-added services license and got started."

By September 1994, Hur had created iNet, the first private dial-up ISP in Korea. It primarily catered to the business sector, instead of the residential sector, as the cost of access remained far beyond the reach of most Koreans. By the following year Hur had raised nearly $5 million, with an additional $8 million credit line, mainly from TriGem, which figured that the ability to get online would be another selling point for TriGem hardware.

One of the main problems with dial-up access, however, was that as companies were making a phone call with their computer, they had to pay per minute. In Korea, as is true with most of the rest of the world, calls—local or otherwise—are charged per minute. So, while Korea had commercial e-mail and Internet service in the early 1990s, its users largely remained businesses that were willing to shell out hundreds of dollars per month to have access.

By contrast, in the United States local calls have been unlimited at a flat rate for many decades. So, when local dial-up Internet arrived in the 1990s, it may have been slow, but it was still available at a flat rate.

As iNet was rapidly rising, KEPCO, the Korean state-owned power company, began to see the value in promoting increased Internet access and theorized that it might be able to use existing infrastructure to become an ISP. During the 1980s, KEPCO had begun constructing a nationwide network of electrical infrastructure, including both transmission towers and overhead and underground carrier lines. The company included fiber optic cable in the underground lines for the purposes of communication and control, to manage Korea's electric power grid.[81]

Fiber optic cable has a very high data capacity, so it was thought at the time that this would be more than the state could actually use for its own communications purposes. Further, in the early 1990s the infrastructure was laid for a nationwide cable television network, giving every home access to cable television. The network was designed to be bi-directional, enabling each home to not only receive transmissions, but to send as well. Thus, the fiber optic cable became the backbone of the nationwide Internet, and the cable television infrastructure became the solution to the "last mile" problem, which postulates that it remains easy to lay down long-distance network infrastructure, but not as easy to transfer that data connection to each and every remaining home and business, particularly in rural and remote areas.

As commercial Internet access began to slowly grow in Korea by the early 1990s, KEPCO wanted to use its existing lines to become an ISP. However, like anywhere else, there were politics to overcome first. The Ministry of Industry controlled KEPCO, and so, naturally, when the Ministry of Communications caught wind of what KEPCO was trying to do, it felt that KEPCO was overstepping its bounds. In other words, the Ministry of Communications felt that it should be in charge of all data communications-related activities, and not cede this to another ministry.

In 1996, under President Kim Young-sam's administration, the government decided that in order to resolve this situation it would create a new company, Thrunet. Under the control of the Ministry of Communications, Thrunet would use KEPCO's lines. The government sought investors from existing Korean companies. TriGem, as one of the largest Korean IT companies, invested about 10 percent of the entire amount. With TriGem as the largest single donor, it only made sense to have the company's head, Lee Yong-Teh, serve as the president of Thrunet.

Around the same time, a California startup called @Home was beginning to offer a similar cable modem service across the United States, making high-speed Internet access available for a flat fee, well within reach of many middle-class families at about $50 per month.[82] Chon Kilnam, who followed much of the tech news from across the Pacific, quickly recognized it as a model for Korea to follow, even if initially the service didn't have much of an impact in the United States.

Chon knew that the cheaper, flat-rate dial-up connection in the United States was preventing many American users from switching to cable, and he also knew that the same flat rate would be the "killer app," the main reason that Korean subscribers would turn to Thrunet. He supposed that for most people the fact that broadband inherently came with more bandwidth was simply a bonus. Chon told his old colleague Lee about this revelation, and Lee agreed.

As head of Thrunet, Lee quickly hired Park Hyunje, one of Chon's best students, to implement Lee's new plan of providing high-speed Internet by using this cable infrastructure. Park, after all, had been network manager of the SDN, and was one of the best-trained and most experienced network engineers in Korea. Park agreed that the way to go after this new broadband market was in fact to simply out-price dial-up. By 1998, Park estimated that 3 million people—about 6 to 7 percent of the population—were using dial-up. Each person who used the Internet for one hour a day paid about $50 per month for the privilege. So it would be up to Thrunet to price its initial offering of one megabit per second at that same level.

"The first users should be power users," Park explained. "In the first three years, we got 1 million customers. [We had been] aiming for 300,000 users."[83]

The continuing Asian financial crisis raised iNet's operating costs dramatically—the exchange rate to the U.S. dollar doubled very rapidly and bank interest rates quadrupled in a matter of months. Hur was forced to sell the company by 1998. While iNet may not have lasted as a company, it served as a catalyst for the industry, setting the stage for the cutthroat competition that would soon make Korea the most wired country on Earth.

Within a year of Thrunet's initial cable Internet service in 1998, its two major rivals, Hanaro (another government-founded startup) and the telephone incumbent KT (Korea Telecom), both began offering a similar level of broadband service (DSL) for $40, and later $25. By May 1999 Thrunet had 63 percent of the broadband market, Hanaro had 35 percent, and KT only 2 percent.[84] Two

years later, Thrunet was hit hard as KT rapidly overtook it, with better marketing and faster speeds. Competition was fierce, recalled Lee Yong-Teh.

"The three companies began competing seriously," Lee said.[85] "That's why
Korea became the number-one broadband Internet service country, was
because of these two giants—and Thrunet went bankrupt."

For its part, the Korean government acted to further the expansion of
Internet access and education even in light of severe economic troubles. It continued the two-year-old program "Framework Act on Informatization Promotion," an ambitious public and private multi-year investment plan to enact
1995's initial "Korea Information Infrastructure" strategy.[86] KII, as it came to be
known, contained a three-prong public and private initiative to expand IT
capacity in Korea. The idea was to expand a government-side backbone network, and ultimately bring more and faster access to the masses.

In fact, when Kim Dae-jung was elected president of Korea in mid-1997, the
government promoted the wiring of apartment buildings with a "Cyber Building
Certificate," which assigned a ranking of first through third class, depending on
how fast the building's network capacity was.[87] As such, new apartment buildings could easily distinguish themselves by ranking. Further, with more than half
of Koreans living in dense urban apartment buildings, this implementation
made it easier for broadband access to reach more people more quickly. To top
it off, during his 1997 inauguration speech President Kim pledged to make Korea
the most "computer literate nation in the world."[88]

Within his first term, President Kim implemented two Internet-related government programs, "Cyber Korea 21" (1999) and "Ten Million People Internet
Education" (2000). The former aimed to both expand Korean e-government
services, digital literacy, and e-commerce, while the latter was a public education campaign designed mainly to teach Korean homemakers how to use the
Internet. The government issued subsidies to private Internet and computer
training facilities to lower the cost of such courses. This project targeting non-
employed women was particularly interesting, as authorities realized that in
order to get more and more Koreans to use the Internet at home, they needed to
convince the main member of the family that controlled the household finances,
typically mothers. These mothers felt that they could support their children's
education by becoming more aware of what the technologies were capable of.
To set a public example of how the Internet was being used in government,
President Kim sent the first e-mail to his presidential cabinet in February 2000.[89]

"Technology Like It's Water from a Water Fountain": eSports, Citizen Journalism, and Urban Design (1996–Present)

In 1990, just as the old guard of students like Park and Hur were leaving, Chon
Kilnam brought in a new graduate student at KAIST, Song Jae-kyeong, also

known as Jake Song. Song was initially impressed by Chon's interest in making his students both worldly and nationalistic. Chon made sure that his students were exposed to Western business and academic norms, but also made sure that their work was benefitting Korea. Initially, Song took on a project of building the first Korean-language text editor.

But it wasn't long before the young twenty-seven-year-old doctoral student started to get sidetracked from his main project. By 1992, some of the older students at KAIST had developed KIT MUD, their own Korean-language MUD (multi-user dungeon), a text-only fantasy game that had been popular in the 1970s and 1980s in American computer science departments. The program's lack of graphics wasn't a big deal for students playing on slow networks, who had sporadic access to computers—but with knockoff versions of eight-bit computers and gaming consoles starting to trickle into Korea, Song thought he could make a better, graphical version of this game.

He also observed firsthand Chon's dual nature of relaxing certain cultural norms—like allowing his students to smoke in front of him—but also imposing a strict regime of organization and precision. Chon required every student's one-page briefs to include a serial number for archival purposes, like Jake91–001. "A document without a serial number was not a document," Song recalled, "He would throw it away."[90]

As the months went on, Song started to ignore his course work and school responsibilities more and more, until finally in mid-1993 he dropped out of the doctoral program and instead got a job working for a new Korean word processing company. While there, he again spent his free time working on his new massive multiplayer online role-playing game, or MMORPG. By 1996, he had released his first video game title: Nexus: Kingdom of the Winds, a similar fantasy game loosely based on Korean medieval history. That same year, Korea's fist Internet café, NET, opened in Seoul. NET and other Internet cafés like it helped more and more people discover these games. But it was Song's second title, Lineage (released in February 1998), that became a huge success. The release of this new game marked a substantial turning point in Korean gaming history and led directly to the creation of the Korean eSports league.

Lineage, based on a South Korea comic book of the same name, was a medieval-era fantasy simulation game, in which players could participate as an Elf, Dark Elf, Knight, Prince, or Magician. Players could also join with others to form a "blood pledge," or clan. The main reason why Lineage was so popular was that the players could interact with each other in teams of dozens, or even hundreds at a time. Both Nexus and Lineage were among the earliest and most popular MMORPGs in the world, reaching many thousands of players online at the same time.

For the next several years, Lineage was the world's most popular subscription-based MMORPG, reaching 2 million players, until the release of Blizzard

Entertainment's most popular MMORPG, World of Warcraft, in 2004. (Four years later, World of Warcraft had reached 10 million subscribers.)

However, shortly after the release of Lineage, on March 31, 1998, Blizzard Entertainment released StarCraft. Within the game's first year, it sold 1.5 million copies worldwide, a large portion of which was in Korea. Today, the game has sold over 9.5 million copies, nearly half of which are in Korea. Thousands of Korean teenagers and twenty-somethings (most of them boys) instantly began to spend hours and hours engaging one another in battle.[91]

While the Korean Internet was booming just as fast as any other, economic realities put the brakes on many of these new startups. In the fall of 1997, Korea and many other Southeast Asian countries experienced what came to be known as the Asian Financial Crisis. The troubles began in July when Thailand decided to float the value of its currency, the baht, and lift its peg to the American dollar. This led many investors to pull their money out of Thailand as they lost confidence in the currency. As investors pulled their money out, more and more companies started having significant capital problems as well. Also in July, Korea's third-largest automobile manufacturer, Kia Motors, asked the government for emergency loans and was put under governmental bankruptcy protection. Three months later, the government took over the company. By late December, Moody's lowered the credit ratings of many major Korean companies, including Pohang Iron and Steel Company, the Korea Electric Power Corporation, SK Telecom, Hyundai Motor, and Samsung Electronics, and later further lowered Korean state bonds to junk status.[92]

By 1998, as the Asian Financial Crisis continued to engulf the entire country, many middle-class entrepreneurs with access to some small amount of capital decided to open PC Bahngs, as they didn't require huge amounts of investment—less than $100,000—and office and retail real estate were cheap to come by. That same year, Internet cafés gave way to PC Bahngs as more and more customers wanted to use this environment to play games like Nexus, Lineage, and, increasingly, StarCraft. While nearly all PC Bahng patrons had computers at home and would use them for educational or productive purposes, these new "third spaces" outside of school/work and home created an environment where kids could be around other people and take on virtual enemies. Plus, on a slower connection at home, a game like StarCraft starts to "lag," and this delay of sending and receiving information becomes frustrating for serious players. The faster the connection is, the faster and more fluid the gameplay.[93]

Rapid demand to play these games on a high-speed Internet connection led to a massive explosion in PC Bahngs, which were quicker to bring in faster connections than most residential homes. These entrepreneurs enticed customers with subscriptions and promotions—the more you play, the less you pay per hour—and the PC Bahngs would host regular tournaments between clans as a way to keep things busy. By the end of 1999, there were over 15,000 PC Bahngs in the entire country, tripling in number from the previous year. PC Bahngs also contributed significantly to the rise of Korea's computer gaming industry: in 1999, national sales of online games reached about $16 million, and by the following year they skyrocketed to $148 million.[94]

Beyond the Asian Financial Crisis of 1997, the administration of President Kim Dae-jung was beset by the challenge of dealing with Korea's greatest issue: North Korea. Since the conclusion of the Korean War in 1953, the peninsula has been divided between north and south—technically a peace treaty has never been signed. North Korea remains a communist dictatorship and has little contact with the outside world. Both sides have heavily armed the border along the demilitarized zone that splits the peninsula in two. During the 1970s and 1980s, at the height of the Cold War, South Korean military and political leaders prided themselves on their anti-communist stance. One legacy of this is that nearly all North Korean-related websites are banned in South Korea, which remains the main example of online censorship in the south. North Korea, of course, remains offline, except for a very small handful of party elites, foreign diplomats, and others.

During his presidential campaign, Kim proposed what came to be known as the "Sunshine Policy" of engagement toward the country's estranged other half. The idea was that through peaceful cooperation in the short term, the differences could eventually be reconciled and Korea could once again become unified in the long term. Two unprecedented meetings between President Kim Jong-il and Kim Dae-jung were held in Pyongyang in 2000, an act of mediation and peace that earned the South Korean president the Nobel Peace Prize later that year.[95]

However, shortly before the Pyongyang summit, Korea had a national legislative election in April 2000. During that election, Roh Moo-hyun lost his bid for a parliamentary seat representing the southern port city of Busan. Although he had been active in politics for the previous decade, he had not won an election since 1988. While a minister of Parliament in 1988, Roh was part of parliamentary hearings investigating Korea's military dictatorship between 1981 and 1987, and his rallying against corruption and abuse of power bolstered his public image as a scrappy public servant fighting against the old guard.[96] The fifty-three-year-old, with his compelling personal story, campaigned on the promise of a Korea not plagued by regional divisions that had traditionally divided the country.

Despite Roh's loss, he drew support from thousands of Koreans who, in unprecedented numbers, flocked to his official website to leave messages of support and encouragement. Two days later, on April 15, 2000, one supporter, Lee Jung-Ki, left a message suggesting someone create a fan club site for Roh—called Rohsamo (or Nosamo)—taking its name from the Korean acronym "People who love Roh." While today, politicians worldwide organize around Facebook pages and YouTube channels, this was the first time in the world that such an organized campaign erupted online. Within the first ten days, the club drew more than three hundred members, ranging in age from teenagers to elderly citizens in their sixties.[97]

In less than a month, local branches of the fan club sprung up all over Korea, paving the way for the launch of the official website on May 17, 2000. On June 6, 2000, the first national meeting of Rohsamo was held in Daejeon, a city about two hours south of Seoul. But more important than the choice of city was the choice of venue: a PC Bahng.

OhmyNews, a participatory "citizen journalism" website that had just launched in February 2000, devoted live coverage to the launch of the Rohsamo group at the Daejeon PC Bahng. Korean journalist Oh Yeon Ho, who wanted to counteract the dominance of the traditional conservative Korean-language media, had created OhmyNews. On the advice of a journalist friend, Oh had left Korea in 1995 for Regent University in Virginia—a conservative Christian school founded by evangelist leader Pat Robertson. Oh has repeatedly said that while he does not share the political views of Regent's founder, he does admire Robertson and his allies' ability to mobilize and create media supportive to their cause.[98] Oh graduated with a master of arts degree (with a focus in public writing) in 1997.

Oh says that a journalist friend told him that to know America, you have to know how the American conservative right operates.[99] Part of Robertson's method was to start his own media outlets, as a way to further his own religious, social, and political goals. Robertson's empire grew by purchasing television stations in the 1960s and 1970s, and creating the Christian Broadcasting Network. Today, Robertson's personal net worth is measured in the hundreds of millions of dollars. While Oh was not after the same level of personal wealth, he was intrigued by Robertson's aggressive approach to establishing his own media outlet. But instead of television, he would use the Internet.

Upon his return to Korea, Oh began laying the groundwork for a concept that he had first conceived of while a student at Regent—that is, to tap the collective wisdom and power of average people on the Internet who were not represented in the traditional media. In 2000, OhmyNews was born. Its name is a play both on his family name, Oh, and on the English expression "Oh my God!" which around that time had become a popular English catch phrase in Korea.

OhmyNews promised to alter the fundamental premise of traditional print media, in which the reporters wrote and the audience read, without any kind of interaction or mediation between the two. Its motto declared, "Every Citizen Is a Reporter." Instead of relying on information fed to them, readers could become "citizen journalists" and would be able to be both reporter and reader at the same time, contributing original content about topics that were happening in their own small part of the world.

The site became the most well-known and successful example of the growing phenomenon that attempted to merge trained professional journalists with an army of low-paid amateurs. It began with a staff of four professional reporters and editors and seven hundred citizen reporters covering both domestic and international news of interest to a Korean audience. OhmyNews currently has a staff of fifty-five trained reporters and editors, who contribute approximately 20 percent of the site's content. The rest is written by a nearly volunteer staff of more than 50,000 Koreans who are usually paid between $2 and $20 for their work—the legions of citizen reporters seem to be motivated by the simple act of being published and being read, rather than monetary gain. The citizen reporters, most of whom are men between the ages of twenty and forty, usually write softer news—features, reviews, or opinion pieces—rather than hard, breaking news. Each time a story is written and submitted to the site, it goes through an editing and fact-checking procedure, and will not be published until it has been approved. However, not every story is approved—as much as 30 percent of the two hundred stories submitted daily are rejected.[100]

OhmyNews has become very successful in Korea. In a 2003 survey by the Korean magazine *Sisa Journal,* OhmyNews was ranked as the sixth most influential Korean media outlet.[101] It was the only online outlet to make the top ten.

OhmyNews expanded to include an English (or "International") edition in 2004 and a Japanese edition in 2006.[102] Despite the fact that OhmyNews Japan began with $11 million of funding from SoftBank—one of Japan's largest mobile phone companies—it closed down in July 2008. The International edition takes a global perspective, while the Japan edition focused mainly on Japan and international news of interest to Japanese readers.

Since the creation of OhmyNews, there have been a number of citizen journalism projects begun in the United States, including Backfence, NewAssignment.net, and Assignment Zero, but many of them have started and fizzled for lack of interest and/or funding. At least on the American side of the Pacific, citizen journalism has been more hype than success—it has yet to catch on in a meaningful way that rivals traditional media outlets like OhmyNews does in Korea.

Put more succinctly, as Nicholas Lemann wrote in *The New Yorker* on August 7, 2006:

> In other words, the content of most citizen journalism will be familiar to anybody who has ever read a church or community newsletter—it's heartwarming and it probably adds to the store of good things in the world, but it does not mount the collective challenge to power which the traditional media are supposedly too timid to take up.[103]

OhmyNews, by contrast, is more successful in Korea largely because the existing media establishment is largely singularly conservative and does not adequately provide alternative voices. Further, OhmyNews attempts to hold government accountable in ways that the traditional media has not typically done. One of the site's shining moments came in late 2002, when OhmyNews got its big break as both the creator and the reporter of news.

On June 13, 2002, two American sergeants driving an armored vehicle during a training exercise accidentally killed two Korean middle school girls. The girls had been walking on the side of a road outside of a town near the DMZ when the vehicle struck them. After the accident, anti-American sentiments surprisingly reached an all-time high in Korea, with many arguing that the longstanding, unshakeable military alliance between the United States and South Korea was unfair to Koreans. Korean society became so tense that a few restaurants in Seoul displayed signs that read: "Americans are not welcome here." This statement was shocking for a country that owes its freedom to American political, economic, and military support.[104]

By late November, the soldiers, under the U.S.-South Korea Status of Forces Agreement, were tried in American military courts and were both acquitted of negligent homicide. Not surprisingly, this decision only worsened the already tense situation between the two countries. On November 27, 2002, a Korean television station broadcast a documentary about the trials and their outcomes, and in the early hours of the next day an OhmyNews reporter writing under the name AngMA (real name: Kim Ki Bo) posted a message on several Internet forums, including OhmyNews, calling for a nightly candlelight vigil and protests near the U.S. embassy in Seoul on November 30. The first night, 15,000 people showed up. Two weeks later, the crowd swelled to 100,000.[105]

While OhmyNews ran story after story about the anti-American protests, the traditional conservative press, including *Chosun Ilbo*, Korea's largest national daily newspaper, devoted far less ink to the original deaths, the military trials, and the ensuing protests. One media study later found that in the ten weeks immediately following the accident, OhmyNews published 1,010 articles on the subject, while *Chosun Ilbo* published less than one hundred. From the late November verdicts until the end of 2002, OhmyNews published 1,965 articles relating to the case, and *Chosun Ilbo* only 604.[106]

AROUND THE SAME TIME that Rohsamo, OhmyNews, and other media and political advances were under way, there were efforts between the government and private industry to try to formalize the countless number of amateur tournaments and games for Lineage and StarCraft. At the time, PC Bahngs across Korea were filled with many teenagers and twenty-somethings who wiled away their time attacking, strategizing, and killing each other in a virtual world—but it never went beyond the confines of a single PC Bahng, or Blizzard Entertainment's Battle.net online rankings.

As a way to promote more domestic organization and competition, the Korean eSports Players Association (KESPA) was born in 1999, in conjunction with the Ministry of Culture, Sports, and Tourism as well as sponsorships from large Korean companies. KESPA formalized the concept of a "professional gamer," with a registration and acceptance procedure for all would-be players on nine different "official" computer games. Today there are nearly five hundred officially registered "pro gamers," nearly all young men, most of whom are aged seventeen to twenty-four, who make tens of thousands of dollars per year—the top ten players make $100,000 or more. One of the best and most popular players (he has over half a million members of his fan club), Lim Yo-hwan, earned over $300,000 in 2005.[107]

Of these nine official KESPA games, StarCraft remains the most popular, and the most organized. In Korea's StarCraft Pro League—the largest and most developed such league in the world—eleven corporate-sponsored teams and one Air Force team (all based in Seoul) compete for the annual nationwide StarCraft championship title. Each team competes in twice-weekly games in head-to-head and two-on-two match-ups.

While the idea of paying kids such large amounts of money to play video games may seem somewhat ridiculous, Daniel Lee, former coach of the eSTRO team, believes that it's no crazier than professional golf or football. "Just like any other [professional] sport," he says, "it's very systematic and it does take athleticism and it deserves much more recognition than it has right now. We're turning the game into an art form."[108]

Lee, who coached StarCraft in Korea for nearly a decade, points out that playing a video game this well requires an incredible amount of dexterity, hand-eye coordination, speed, and strategy. "In pro-level StarCraft, the average actions per minute is three hundred—it's like playing the piano at a very high level," he says.

As a coach, Lee imposed a very strict schedule for his players. All of them lived together in dorms for several months out of the year, where their schedules and daily lives were planned, hour-by-hour, military-style. During his tenure, eSTRO players would wake up at 9 A.M. and have two hours to change, shower, and eat a breakfast prepared by the team's professional cooks. From 11 A.M. until 4 P.M., the team practiced, mostly playing against one another, testing out new strategies and reviewing footage of games that they previously lost. At 4 P.M., the team took a break to eat, followed by a physical workout at a gym. Gaming practice would resume at 7 P.M., followed by a "light supper" at 10 P.M., and then two more hours of practice until bed at 1 A.M.

Today, the size of the Korean eSports industry is over $80 million collectively, and shows no signs of slowing down. Marketing and endorsement deals feature various pro gamers on Korean credit and bankcards, teddy bears wearing uniforms of each team, as well as branded computer accessories. Most strangely of all, there's even an "official" dentist of the Pro League—the White Style Dental Clinic in Seoul.[109]

WHILE OHMYNEWS, ROHSAMO, and an electrified political generation and hundreds of pro gamers are particularly unique to Korea's early twenty-first-century history, the country also is currently undergoing the most ambitious tech-oriented urban planning project in history: New Songdo City, better known simply as Songdo. The city markets itself as an amalgam of the best that modern technological life has to offer. It is a joint project between Gale International (a major American developer) and POSCO (a major Korean steel firm), and its final cost will be approximately $35 billion, likely making it the largest private real estate venture in human history.[110]

Songdo is also trying to be the first large-scale urban area in the world to build "networking ubiquity" into the very fabric of the city itself.[111] The idea is that a simple ID card—likely through the use of a biometric parameter, such as a fingerprint or a retinal or vein scan—will allow residents to open the door to their apartment, access their workplace, use public computer terminals to order things like groceries or movie tickets, or get on a train. Further, through the predominant use of short-range wireless technologies (RFID) and longer range ones (CDMA, 3G, WiFi), the entire 1,500 acres will be networked within the island and connected to the larger Internet from virtually every point in the city. In industry jargon, this is known as a U-City, with U standing for "ubiquity."

Each resident will be issued a "smart-card" that can be used as a house key, to pay a parking meter, to buy something from a store, to pay a public transit fare, to access the public bicycle sharing network, or for any other number of activities. City planners anticipate scenarios where public recycling bins could use small, wireless RFID tags to send a redemption value directly into the bank account of someone who recycles a soda can, or where pressure-sensitive floors in homes with elderly residents could automatically alert medical authorities in the event of a fall.

"[We] want to have technology like it's water from a water fountain," says Carl Seaholm, one of New Songdo's board members. "You'll be able to walk up to computer kiosks anyplace and let it know who you are. You'll have access to goods and services—we're putting in all the pieces in place to have that happen. Maybe school kids will be able to pull up a list of their homework assignments— the security systems will be interlinked. [You'll be able to monitor] what the inside of your apartment looks like. Shopping will be on the Internet, but it will be local, as you'll be linked into your own home portal. Basically you end up with a very free-flowing information environment."[112]

Once completed in 2015, New Songdo City is expected to have 65,000 permanent residents and 300,000 daily workers who live and play in a massive 1,500-acre area. While it will be a Korean city, it will also be the most foreign city on the peninsula. English is expected to be the lingua franca (like in Hong Kong or Dubai), as half of its residents will be non-Korean. The island's hospitals will be operated by Johns Hopkins University, while some Harvard University education faculty are expected to run its schools.[113]

Further, New Songdo's urban design is inspired by major cities around the world—with a New York-style Central Park, Battery Park, and Rockefeller Center,

Venetian-style canals complete with water taxis, dense urban areas à la London or Paris, garden districts like Savannah, and a Sydney-like iconic waterfront cultural center—and will have 40 percent green space, significantly higher than most other Asian cities.[114]

The project is made possible through bureaucratic support from the Korean government, including tax and land-use policies that normally favor Korean companies. Because Songdo is also a free economic zone, it does not have the obligatory employment quotas for veterans and the elderly that most other workplaces in Korea require. Further, foreign companies can easily establish their own schools and hospitals, which otherwise would be quite difficult. Finally, as a way to entice major international corporations and investors, the Korean government is allowing foreigners to file documents in English instead of Korean.[115]

Songdo rises out of what once was a landfill island near the shoreline along the East China Sea. It was a desolate, empty piece of territory before undergoing this transformation of urban renewal in the largest sense possible. Today, many of the office and residential towers have been completed, but the entire technological infrastructure has not been finished yet.

An account by a visiting Philippine journalist in November 2009 put it this way:

> We were . . . shown how the system also could wirelessly and remotely control all the residential units' electronics, appliances and security. Songdo residents will be constantly connected with everything and everyone they need to deal or socialize with without having to expend fuel and energy by driving there.
>
> If you do need to go anywhere in the city, the road system is wonderfully maintained and rational. The authorities also built a subway extension to old Incheon and even to Seoul. This road and subway infrastructure was built ahead of any development. Also built ahead of anything was the complete infrastructure for power, sewage treatment and drainage. Easements on all drainage canals and channels are between 50 and 100 meters wide—compared to the measly five-meter easements we have here in Metro Manila![116]

Songdo is located in the upper northwest corner of South Korea; this is also the site that was invaded by General Douglas MacArthur during the Korean War. It is adjacent to the city of Incheon and is now connected via a seven-mile-long bridge to Seoul Incheon International Airport and will eventually join Seoul's subway network. It sits just a handful of kilometers south of the DMZ.

Songdo is essentially an extreme manifestation of what "futuristic" cities should be like: a vast, safe, planned city where high-speed wireless technologies are what make the city tick. This connectivity is the city's lifeblood: it is what opens doors, transfers money, and keeps track of natural resources. While New Songdo City sounds impressive on its face, it is still unclear how all of these technologies will be laid out, or how much of a premium its residents will pay for them, given that nothing like this has ever been attempted anywhere else in the world. Only Ülemiste City in Tallinn, Estonia, is similar, albeit on a much smaller scale, and does not have the residential component that New Songdo is planning. Further, as an April 2005 article in the journal *The Next American City* pointed out, there are many levels besides white-collar work and play locations that Gale and the planners of New Songdo seem to have overlooked:

> In adhering to the formula that good design will make a city commer-
> cially appealing to corporations, the developers seem to have glossed
> over the people who aren't part of their business plan—namely, the poor
> and working-class. It is impossible to have a city where only white-collar
> office workers live because office functions need support industries: they
> need people to launder their shirts, print their documents and make
> their take-out dinners. Those in the service industry will necessarily
> make less money than their counterparts and live in less desirable
> neighborhoods; how does one plan for this?[117]

The Dark Side of the Korean Internet: Online Addiction, Cyber-Humiliation, and the "Wisdom" of the Mob (2001–Present)

While the combination of Korea's history, culture, and innovative spirit has been able to produce some of the world's most amazing technological outcomes, it has also produced three tragic ones: online addiction, cyber-humiliation, and mob rule. As more and more citizens come online, especially in the developing world, it is important to remember how cultural and social norms can mix with technology in volatile ways. Just as Korea may represent a positive, wired future, it may also provide a warning of some of that future's more undesirable possibilities as well.

Because of its highly developed Internet infrastructure and hugely popular online gaming culture (mostly StarCraft and Lineage), it's no surprise that Korea was where the first deaths resulting from computer gaming addictions were reported anywhere in the world. From 2001 to 2004, there were two known gaming-related deaths in the country, but that number spiked to ten in 2005.[118] The actual cause of death in many of these cases was largely preventable—a disruption in blood circulation caused by sitting in a single restricting position for too long at a time.

In 2002, a man died in Kwangju after playing for three and a half days straight. In August 2005, in an even more extreme case that made international headlines, a man from the southern city of Daegu collapsed and later died in a PC Bahng after playing StarCraft for nearly fifty hours, with only minimal bathroom, sleeping, and food breaks. Authorities found the cause of death to be "heart failure stemming from exhaustion." In December 2005 a thirty-eight-year-old worker suffered a similar fate after averaging nearly twenty-one hours of game play per day for twenty days. There have been other reports in Korea of real-world monetary theft and of prostitution to earn money for a game, and even of physical violence stemming from a game-related dispute.[119]

Since 2002, the Korean government has spent large amounts of tax money to try to study and combat Internet and online gaming addiction. There are now public awareness campaigns, a gaming addiction hotline, and no less than 140 centers for Internet addiction nationwide. The country has more than a thousand counselors trained in the treatment of Internet addiction.[120] In 2003, an online addiction scale—known as the K-Scale (for Korea)—was developed to accurately evaluate the state of someone's addiction. By 2006, the government had created a related scale specifically for gaming addiction.

"The definition is that if they use the Internet too much, then they have withdrawal," says Kim Hyesoo, who from 2005 to 2007 headed the Net Addiction Treatment Center in Seoul. "Symptoms of withdrawal [can make people] very anxious to use the Internet and to use online games, that kind of thing. It disrupts their life—they don't want to make a friend or to talk to a real person in an offline life. They cannot eat regularly, they cannot sleep regularly, that kind of thing."

Kim, who has since gone into private practice, says that while the number of hours online can be an indicator of Internet addiction, it has more to do with the mental state and the way that kids feel when they spend time offline.[121] In therapy, kids are encouraged to spend their time engaged in much more social outdoor activities, like sports, camping, or even other interactive board games, like chess.

According to the Korean government's own estimates, potential for addiction is quite high. In April 2009, the president of the Korea Agency for Digital Opportunity and Promotion (KADO), which runs the Center for Internet Addiction Prevention and Counseling, estimated that as much as 5 percent of Korea's population (around 7 million people) have an "unhealthy attachment to the Web." Moreover, according to a November 2007 article in the *New York Times,* "up to 30 percent of South Koreans under 18, or about 2.4 million people, are at risk of Internet addiction."[122]

While the real number of actual addicts is hard to measure, there has been some data made available in recent years. In October 2008, the Korean Ministry of Education released the results of a study of 5.2 million Korean schoolchildren, which found that nearly 2 percent of them were addicted to the Internet—defined as spending four hours or more online per day, and reporting associated problems in their daily lives. From January through July of 2005, there were over 7,600 government-sponsored counseling sessions for game addiction, which was on track to beat the previous year's record of almost 9,000 in 2004. The previous year, 2003, had just over 2,200 sessions. However, Kim Hyesoo says more people are coming forward for treatment than in the past due to the extreme examples of deaths and other real-world consequences—which is a good thing. Further, because of government prevention campaigns and increased public awareness, Kim adds that since 2005 there has been a decrease in the number of Internet addiction cases.[123]

Kim says that most kids are cured after three to six months of treatment, which usually involves a diminishing of extended periods of game play. That time is typically replaced with another less obsessive activity, such as basketball or chess. In more extreme examples, Korean boys are sent to the Jump Up Internet Rescue School, which opened in 2007.[124] The school is a twelve-day boot camp in a forest region one hour north of Seoul, where the boys participate in a number of summer camp-style activities, including pottery, obstacle courses, and therapy sessions as a way to alleviate their obsessions with online gaming, which in these rare cases can often dangerously approach twenty hours per day.

In April 2010, the Korean government began testing a policy to bar school-children from gaming between midnight and 8 A.M. and introduced another policy to slow a gamer's connection after they have been connected for several hours. While gaming's primary demographic is young men, there are plenty of obsessed adults as well: one Korean couple was arrested in March 2010 after starving their real-life three-month-old baby to death in favor of taking care of a virtual child.[125]

While gaming has not reached this level of obsession in the United States, the data that has been collected on Korea's Internet- and game-obsessed kids is often used as a reference for Internet addiction and is used to bolster arguments as to why more needs to be done to combat this issue. One Oregon physician, Dr. Jerald M. Block, submitted an editorial to the American Psychiatric Association in March 2008 arguing to have Internet addiction included in the next edition of the *Diagnostic and Statistical Manual of Mental Disorders* (DSM-V), due out in 2012. In it, he based his argument for inclusion in the DSM-V largely on data available from papers and cases about Korean Internet addicts.[126] However, despite all the attention, there is no official diagnosis for Internet addiction, and many criticize the term "addiction" as being too strong. Still, the fact that Korea (and to an increasing extent, China) has devoted considerable resources to this issue suggests that either Korea is responding to a problem that most of the rest of the developed world doesn't have, or that the rest of the developed world may be under-diagnosing the problem.

If one of the outcomes of being an extremely wired country is, in fact, Internet addiction, another is the potential for public humiliation suddenly accelerated through cyberspace. On an otherwise very average day in June 2005, a woman in her twenties boarded a Seoul subway car accompanied by her small dog. While on the subway, the woman's dog defecated on the floor of the train car. The woman seemed to not react in any way, nor did she attempt to clean up the mess. Other riders asked the girl to clean up the mess, and when she didn't, a nearby woman on the train offered her a tissue. When the girl still didn't respond at all, another woman on the train snapped a few photos on her cell phone. Later that day, those photos were uploaded to a popular Korean website, where the girl was quickly nicknamed *gae-ttong-nyue* (Dog Shit Girl, or Dog Poop Girl). Within days, the girl's identity was determined based on the type of bag she was carrying and the type of the watch she was wearing. Online vigilantes published her personal information, and the Korean Internet was brimming with requests for information about her parents and other relatives, likely for further embarrassment and/or harassment. The topic even found its way into Sunday sermons in Korean churches in the Washington, D.C., area. Soon after, apparently the girl was so ashamed that she ended up making an online public apology and quit her university studies.[127]

Given the speed with which this girl's privacy was invaded over a relatively minor transgression, it's not so surprising that the incident happened in Korea. After all, this public shaming is the result of using the collective "wisdom" of ordinary people who were committing acts of journalism by documenting a small, but offensive, incident and who were then disseminating it on the largest communications network ever built. The scandal grew in a way that could not have happened in many other countries. Further, the small social network of the country (50 million people) meant that she could be tracked down with much greater ease than in a much larger country, like the United States.

Still, once that nugget of offensive information was released—the series of photos—the girl instantly became a target without bounds, defense, or recourse. In the seeming anonymity of the Internet, anyone can hide behind pseudonyms and scream insults, or even invade someone's privacy with very little restraint. While many of the commentators say that she was asking for it by allowing her dog to poop on the subway to begin with, there's a large difference between hurling criticisms or insults and encouraging others to harass her family.

And while this girl may only have been famous on the Internet for fifteen minutes, the larger issue remains of how to deal with online groupthink, particularly in the age of rapid dissemination of text, photos, and videos online. Worse still, those media files can be stored indefinitely on Google, the Internet Archive, or countless other sites. For any minor transgression that one commits, it's possible that this mistake will be committed to a permanent digital memory. As Daniel Solove, a privacy expert at the George Washington University

Law School, writes, the Internet can take such a strange incident and sear it into our collective memory:

> She will not be forgotten. That's what the Internet changes. Whereas before, she is merely remembered by a few as just some woman who wouldn't clean up dog poop, now her image and identity are eternally preserved in electrons. Forever, she will be the "dog-shit-girl"; forever, she will be captured in Google's unforgiving memory; and forever, she will be in the digital doghouse for being rude and inconsiderate.
>
> Consider the famous incident involving the "Star Wars Kid," a sad tale of a nerdy 15-year kid who filmed himself waving a golf ball retriever around as if it were a lightsaber. To tease him, some other kids digitized it and posted on the Internet along with his name. It was downloaded by millions around the world, and new versions of it quickly emerged replete with special effects and music. Forever, this person will be known as the Star Wars Kid. There's even a Wikipedia entry for him![128]

The Dog Poop Girl incident clearly illustrates the unique outcome when a small, wealthy, wired society under the influence of citizen journalism and drunk on pervasive gadgetry meets a social situation that offends it. The indiscretion gets captured and thrown to the wired masses, who chew it up and spit it right back out again.

WHILE CHILDREN'S becoming addicted to the Internet and to online games is unfortunate and online public humiliation is tragic, there was one incident in 2008 that tops all Internet-inspired activity in Korea: one that can influence foreign policy.

In late 2007, Lee Myung-bak of the conservative Grand National Party was elected as South Korea's tenth president. Shortly after his inauguration in February 2008, Lee promised to shore up the Korean-American alliance by signing a pending free trade agreement with the United States. One way that Lee attempted to do so was to say on April 11, 2008, that he intended to lift Korea's restrictions on the importation of American beef. Prior to 2003, Korea had been the third-largest importer of American beef in the world. However, in 2003 a small outbreak of mad cow disease caused Korea to ban the importation of American beef.[129]

Within days of Lee's announcement, emails, Korean websites, and text messages began to spread with absurd rumors telling people why they should fear American beef or avoid American-owned restaurants or food outlets, such as 7-Eleven or T.G.I. Friday's. In more extreme examples, rumors said that eating American beef would cause instant death. Other rumors, which were very loosely based on misunderstandings and misinterpretations of a paper by a Korean scientist, led many young citizens to believe that Koreans were somehow genetically predisposed to be susceptible to Creutzfeldt-Jakob disease, the human version of mad cow disease.[130]

However, once the rumors started spreading, no one, not even the author of the scientific paper, could reason with the Korean masses: "By the time his rebuttal was published in the largest daily, the debate on the Internet had shifted to new fears, such as how easily Korean babies might catch bovine spongiform encephalopathy (BSE) from diapers made with material from U.S. cattle."[131]

Further, on April 29, 2008, a Korean news program called *PD Notebook* broadcast a documentary on the "dangers" of importing beef from the United States and included a series of significantly false claims that only reinforced many people's already elevated fears.[132] This paranoia that the American government was somehow involved in trying to keep the Korean people down, or that the Lee government was trying to sell the country out to the Americans, shows that the protest was largely driven by emotions spiraling out of control online, and not any kind of meaningful science.

As the *New York Times* reported at the time:

> To many South Koreans, the beef dispute was not entirely about health or science. Nor is it entirely about economics; U.S. beef is half the price of Korean. Rather, it is the latest test of whether their leaders can resist pressure from superpowers like the United States, even if that pressure is legitimate, as is the case in the beef dispute. South Korea had promised to lift the ban once the World Organization for Animal Health ruled American beef fit for consumption, as it did in May last year.[133]

By early May, these fears that had been swirling online manifested themselves in a very public and powerful way—a public protest and candlelight vigil of a few hundred people, which quickly swelled to tens of thousands by mid-May and hundreds of thousands within weeks. The vigils continued, despite the fact that the Korean mainstream media attempted to quell these rumors with articles debunking the *PD Notebook* program.[134]

As the protests continued into the early summer—and were covered live on OhmyNews's online television division—other blatant falsehoods began hitting the Korean Internet, with one video hoax made by a Korean-American in Los Angeles purporting to show abuses by police officers unleashing water cannons on protestors. In October 2008, a man was jailed for having written and spread false online rumors that Korean police had raped a female protestor.[135]

The protests died down once the Lee government settled on a compromise deal with the United States, whereby Korea would refuse beef from any cattle older than thirty months (it is thought that older cattle are more susceptible to mad cow disease), which is similar to the deal that Japan and Taiwan had previously reached with the United States.[136]

While the protests may be over, the Korean government has made steps to curtail the "netizens'" unfettered access to anonymous commentary. Its tactic aims to prevent rumors from spreading with such credence and rapidity as they did during this series of protests. In fact, shortly after the conclusion of the mad cow protests in Seoul, President Lee addressed an OECD ministerial meeting on information technology held in Seoul on June 15–17, 2008.[137] While he did not directly address the malicious atmosphere created from the rapid spread of news through the Internet, he did lament the "spread of falsehoods" online and expressed his concern that the Internet remain "a space of trust. Otherwise, the force of the Internet could turn out to be venomous rather than beneficial."

Lee's comments were not just idle talk. In late summer and early fall 2008, his government proposed some of the most restrictive Internet-related laws anywhere in the world. The new laws would require that any Internet company, like OhmyNews and the popular Korean portals Daum and Naver, or any other, be subject to the same journalistic regulations as print and broadcast media. Mainly, these laws would require forum and chat room users to verifiably register with their real names, and Korean search companies would have to make their search algorithms public. Further, the Korea Communications Commission would be given the authority to stop publication for a minimum of thirty days on any site found to contain fraudulent or libelous articles. Korea's biggest Internet portal site, Naver, told *The Guardian* that in response, it would cease all of its news production if these bills actually become laws.[138] So far, the government has not made any movement to push the bills any further.

However, it remains unclear exactly how Seoul will be able to police the Korean Internet beyond implementing "Internet etiquette and ethics lessons" in schools. That is to say, how will the government enforce real-name verifiable registration without creating an entirely new bureaucracy and infrastructure for this exact purpose? Further, it would be trivial for Korean-language sites to simply set up shop outside of Korea.[139]

Regardless of whether the law ever goes into effect, the fact that the Internet now has to be regulated in a way illustrates that online access has succeeded in Korea like nowhere else on the globe. In addition to having the high penetration and the highest level of high-speed Internet, the Korean "netizen" population is growing up fast. At least some of them are not yet ready for the level of responsibility that comes with the ability to disseminate information anywhere, anytime.

Bridging Centuries: Building the Next Generation Internet (2009 and beyond)

In some ways, one could argue that the Korean Internet would not have developed as quickly nor as widely were it not for the Japanese occupation between

1905 and 1945. Due to this foreign invasion and occupation of Korea, an entire generation of Koreans that were born and raised outside Korea, and were forced to take Japanese names and speak Japanese, were instilled with a very visceral sense of nationalism. Chon Kilnam was one of these foreign-born Koreans. Despite the fact that he didn't set foot in Korea until 1962 and didn't learn to speak Korean until later in life, Chon retains a deep-rooted passion for a country that he did not live in until he was thirty-six years old.

This love for his ancestral homeland manifested itself in a notion that Korea could better itself, that it could improve, and that it could create technologies few people had ever seen. Chon knew, before anyone else on the peninsula, that Korea was ready to leap forward and was ready to build a new kind of computer networking technology that existed in few other places in the world. Further, he knew that his students could not be like him. Chon knew that Korea would always have enough professors and it didn't need any more—rather, what it needed was creative thinkers, entrepreneurs, people who would take risks and kick-start the technologies that they had worked on in the embryonic stage.

In order to get his students to "be and act like professionals," Chon drew upon his experience studying and working in California. He knew that in order for his plan to work, his students needed to be accustomed to working with colleagues from countries other than Korea. They needed to speak fluent English, and they needed to be used to the idea of taking and archiving minutes at a meeting, as these were all crucial skills that he had learned while abroad. But most importantly, Chon instilled in his students a sense of what it meant to be Korean. And in return, they had something he did not—the experience of being born in the homeland, of speaking the language, of writing *hangul* day in and day out. He insisted over the dinner table that they should appreciate that they were lucky enough to be part of a society that needed them more for their brains in a university laboratory than it did for their bodies alongside the Demilitarized Zone.

In nearly three decades of teaching at KAIST, Chon graduated fewer than twenty Ph.D. students. But many of those twenty have had a significant impact on Korea's technological landscape. Hur Jin Ho started the first private Internet service provider; Park Hyunje managed the first broadband company, Thrunet; Jake Song, who didn't even finish his Ph.D. (Chon insists that he could have if he'd wanted to) used his time at KAIST to be inspired to work on what became the most popular Korean video game ever; Chris Chung, under the tutelage of Chon's mentor, Lee Yong-Teh, conceived of the revolutionary $600 eMachines desktop computer. All remember their time with Chon fondly and instill this creative spirit into their work environments to this day.

IN MAY 2008, CHON retired from KAIST. He is now a visiting professor at Tsinghua University in Beijing and at the Graduate School of Media and

Governance at Keio University in Tokyo. But as nearly a half-century of academia has taught him, there is always more work to be done. In addition to his teaching, Chon's latest project is working as an informal coordinator on the Clean Slate Internet Project, based at Stanford University.[140] This is an extremely ambitious project designed to answer two basic questions about the Internet: "With what we know today, if we were to start again with a clean slate, how would we design a global communications infrastructure?" and "How should the Internet look in fifteen years?"

"The Internet is just like any other engineering system," Chon says. "And like any engineering system, you need a major overhaul once in a while—since it's been forty years, don't you think it's time for a major overhaul?"[141]

What does a major overhaul of the Internet mean? Largely, it means changing the fundamental nature and infrastructure of how the Internet works. Currently the Internet works via a series of rules that were conceived of for a much smaller network in the late 1960s. No one at that time had ever imagined that it would reach anywhere close to 1 billion nodes, let alone the tens of billions of nodes that Chon and many of his colleagues are considering.

In the coming years, Chon hopes that many countries (especially in sub-Saharan Africa) that are largely offline will develop their economies sufficiently to get online. He wants to not only get these countries up to late twentieth-century technological standards, but also bring them into the twenty-first century. He says that even if half of Africa's 1 billion people get online in the next decade, the Internet's current infrastructure is not sufficient to handle that level of growth, much less 10 or 100 billion objects on the Internet. Many gadget and appliance makers already have started to connect household objects, such as coffee makers or refrigerators, to the Internet. Even today, not only are there not enough IP addresses to get all of these people and their gadgets online, the routing tables would need to be reworked in order to make it work.

Further, there are so many problems that exist online today that were never thought of in the days of the ARPANET—unwanted traffic on the network, a denial-of-service attack, even something as basic as spam. However, if the protocols by which e-mail was carried had a built-in authentication system so that one could know who the e-mail's sender was before the mail was received, then the risk of spam would be significantly less. Further, establishing one's identity online as a precursor to accessing webpages or simply using the Internet in any way could cut down significantly on denial of service, or cyberattacks, as was experienced in Estonia in 2007.

Chon thinks of these future changes as being a complete paradigm shift from what the Internet of today looks like.

"Originally, we had roads so people can walk," Chon explains. "Then we had the Industrial Revolution and the invention of the car, and we said that's not good enough, so then we had a paved road, and that improved things tremendously. Then, in the 1950s, the freeway, that really changed the U.S.A. and the world. So then we have two major sorts of the 'clean slate' approach."

However, this long-term research is still in the embryonic stage, and Chon says that he and his colleagues worldwide still don't have a good answer to their most basic questions. Chon says that he doesn't have the technical knowledge as to how to make the Clean Slate Internet actually functional, but he is keeping his role as a coordinator—orchestrating the various components over time to make them come together as one. In this way, he is continuing the work that he began nearly three decades ago.

When he's not thinking about how to fix the Internet or teaching, at the age of sixty-seven, Chon still loves to swim and rock climb.

TIMELINE

March 1, 1943: Chon Kilnam born in Japan

August 14, 1945: World War II ends

June 25, 1950: Korean War begins

July 27, 1953: Korean War ends

April 19, 1960: Student protests in Seoul against Syngman Rhee; 125 students are killed

May 16, 1961: General Park Chung-hee takes power

June 1961: Chon Kilnam visits Korea for the first time

March 1965: Chon Kilnam graduates from Osaka University

December 1967: Chon Kilnam graduates with a master's degree in computer science from UCLA

September 1974: Chon Kilnam completes his doctorate in computer science at UCLA

1974–1976: Chon Kilnam teaches at Los Angeles-area universities

1976: Chon Kilnam works for Jet Propulsion Laboratory; KIET founded

1979: Chon Kilnam returns to Korea; recruited to work for KIET

October 26, 1979: Park Chung-hee assassinated

September 1981: Chon Kilnam proposes System Development Network (SDN)

May 1982: SDN activated between KIET and Seoul National University, 200 miles away, using Internet protocol

September 1982: Chon Kilnam becomes professor at KAIST

March 1983: Hur Jin Ho and Hyunje "HJ" Park begin Ph.D. program at KAIST

December 1983: SDN connected to Usenet

December 1984: SDN connected to CSNET

July 1986: SDN connected to ARPANET; South Korea gets a top-level domain, .kr

February 1990: Hur Jin Ho, Park Hyunje graduate from KAIST

March 1990: SDN connected to Internet using Internet protocol

March 1990: Jake Song begins studies under Dr. Chon at KAIST

September 1994: Hur Jin Ho launches iNet, first private Korean ISP

December 1994: Jake Song quits KAIST, begins working on Lineage full-time

1996: Thrunet founded; Park Hyunje joins; government implements "Cyber Building Certificate"

July–December 1997: Asian Financial Crisis

March 1998: StarCraft released

July 1998: Thrunet offers cable modem service

September 3, 1998: Lineage released

September 1998: eMachines launched

April 1999: Hanaro and KT offer DSL service

February 20, 2000: OhmyNews founded

April 2000: Rohsamo founded

June 13, 2002: Death of two Korean schoolgirls, in Uijongbu

December 19, 2002: Roh Moo-hyun elected president

2005: Yongsan eSports stadium opens in Seoul

June 2005: Dog Poop Girl incident

August 5, 2005: Twenty-eight-year-old Korean man dies in Taegu after fifty hours of StarCraft

May 2006: Construction on New Songdo City begins

May 2008: Chon Kilnam retires from KAIST, becomes a visiting professor in China and Japan

May–June 2008: U.S.-Korea beef protests in Seoul

2

Senegal

Based on our perspective, Senegal is one of the countries in West Africa where such a transition [to an information society] is highly possible. Senior-level officials of the Senegalese government appear to be increasingly concerned about the potential role which information technology could play in the development of society and the economy on the eve of the 21st century.

–U.S. Agency for International Development (USAID), "Senegal Telecommunications Sector Assessment," January 1998

The Cybercafé on Rue Yoff 403

Dakar, Senegal

January 20, 2007

At certain places in Dakar, paved roads disappear into the natural sand like a foot being buried into a slipper. One such street in Yoff—a traditional fishing village that is transforming itself into a middle-class neighborhood—is such a place. Here, upscale gated buildings and fancy clothiers mingle with a small roadside barber's hut. The "salon" is separated from the street by a thin curtain sliding on a simple string.

As is the case in nearly every neighborhood of this incomprehensible city, the gritty but well-stocked "boutiques" flank the street as dependable commercial bulwarks. Coming from the main road, the pragmatically named Route de l'Aéroport, it's very easy to miss the tattered wooden sign in the shape of an arrow nailed six feet up an electrical pole adjacent to one such boutique. This arrow proudly displays, in green paint: "CYBER A 60M," which means that there's a cybercafé about sixty meters down the intersecting street, Rue Yoff 403. The street's name is marked on a French-style blue metal sign. This street sign is so new that hardly any of the locals know the name of their own street— that is, assuming they can read it. After all, only four out of every ten Senegalese are literate in French—the former colonial, and still official language. Far fewer can read or write in their native language, Wolof, Pulaar, Mandinka, or Joola.

Further down Rue Yoff 403 stands a liquor store, where bottles of beer are displayed on sterile white shelves as if they were objets d'art, each standing tall, keeping their distance from their brethren. The store remains empty, and like many things in Senegal, it's not entirely clear why. It could be because it's a quiet and lazy Saturday morning, or because the majority of people in Senegal are Muslim, and thus don't drink. (Or at least their neighbors don't think that they do.)

A few doors past the liquor store sits the cybercafé. Touba Cyber is named for the Senegalese Muslim holy city of Touba, the seat of the Mourides, the indigenous religious brotherhood that exerts a large influence over all economic and political affairs in the country. Touba Cyber might otherwise be easily missed, as there's not much that marks its presence, save one thin wooden white door with a glass window.

Typical of most Senegalese cybercafés, Touba Cyber retains a quiet appearance—they tend to be lit mainly by a little daylight mixed with the pale glow of computer monitors. Most continue throughout the day in near-silence, with the clientele absorbed into their MSN Messenger chat sessions, their e-mails, or European football news. Silent, that is, unless someone forgets to turn down the speakers or put on headphones while blaring a new American rap video from YouTube. A bank of six computers, including one laptop, stands at the ready, each with its own wooden divider.

On this Saturday morning, the only Senegalese customer is Ibrahima Yock, an eighteen-year-old who lives about a five-minute walk away from Touba Cyber. Yock has just finished high school, and is taking a year off to better his English by taking some courses at the University Cheikh Anta Diop in Dakar while he puts together his applications for European, Canadian, and American universities. While this sounds like an admirable and practical plan, his approach to learning about American universities isn't quite laid out with the same rigor. He's typing "Etudier aux USA" (Studying in the USA) into Google and clicking through the various results. He's just stumbled upon the State University of New York at Rockland, where he's begun filling out an online form to "Request Admissions Information."

Yock explains that he comes to this cybercafé at least two to three times during the week and often four or five times on the weekends, particularly when he doesn't have much to do. Back when he started using the Internet in 2001, when getting online required a fifteen-minute bus ride, he'd often have to wait for twenty or thirty minutes just to get online. He'd spend his 500 CFA francs ($1) for a single hour at the cybercafé, where he'd look up music videos of the American rapper Sisqó and check the site of his favorite soccer team, Juventus, of Turin, Italy. Like many Senegalese kids, Yock sought out cybercafés as a cheap form of interactive entertainment.

Of all the schools that Yock would find on this day, why was he looking up information from SUNY Rockland, a small community college in the Hudson River Valley?

"It's in New York," he says in a slight huff, almost scoffing at the notion.[1]

Yock initially doesn't realize that just because a school has New York in its name, that doesn't mean that it's in the city of New York. He seems slightly disappointed after being shown on a map that the school is actually in Suffern, about 55 kilometers (35 miles) from Manhattan.

After an hour of clicking through Google search results in the quiet cybercafé, Yock pays the kid running the manager's table the 250 CFA francs ($0.50) that he's owed. The door bangs quietly behind him as he steps out onto the sandy street, Rue Yoff 403.

TODAY, SCENES LIKE THIS are increasingly common throughout Dakar and the rest of Senegal, where most people don't have computers at home, let alone Internet connections. The easiest way for most Senegalese to get online is to find a cybercafé, where they can share the Internet connection, the computer, and pay only for the time that they need. When even the cheapest used computers cost hundreds of dollars and a home Internet connection costs $50 per month or more—in a country where the GDP per capita (PPP) is $1,700—even a modest investment in computer hardware for a single family represents a significant portion of one's income.

The more expensive cybercafés ($1 or more per hour) are larger, have newer computers, with faster connections, and usually are in more modern buildings with air conditioning and clean floors. The cheaper ones (around $0.50 per hour, sometimes less) tend to be very minimal, and often dusty and poorly lit—like the one on Rue Yoff 403.

Middle- and upper-class people have Internet access at their workplace and sometimes at home: as of October 2008, basic DSL costs about $25 per month plus the cost of a telephone line.[2] Given that this price is comparable to Internet rates in many countries in the global North, significantly fewer people in Senegal can afford Internet access. There are some scattered WiFi hotspots around Dakar and some other cities, at hotels and places of business and government, but there is no established network of WiFi anywhere.

Not surprisingly, Internet access is more common in metropolitan areas, especially in Dakar and its environs. In rural areas, which have sparse electricity and telephone infrastructure, Internet access barely exists. Senegal's nonprofit organization for Internet advocacy, OSIRIS, currently reports that less than 8 percent of the country is online.

An April 2010 government study by the ARTP, the Senegalese communications regulatory authority, found low levels of information and communications metrics even after nearly two decades of high-level policy, reform, and promotion. Among other things, the ARTP reported that just 14.5 percent of Senegalese homes have landlines, while 87.85 percent of homes have mobile phones; that Senegalese homes have just 11.5 computers per 100 homes, while the capital city, Dakar, has 27.5 computers per 100 homes; that of Senegalese Internet users, 83 percent are in Dakar, and that 93 percent of non-Dakar Internet users primarily get online through public cybercafés.[3]

Yet, Senegal's rate of Internet penetration is one of the highest in all of sub-Saharan Africa.[4] Senegal also has the second highest level of international bandwidth among countries below the Sahara (trailing only South Africa), and, as such, can be considered an Internet leader on the continent. Further, Senegal has a relatively high level of education and a stable political and economic system; it has never had any major internal or external conflicts, and is well placed along the West African coastline to benefit from undersea Internet cables. In other words, the Internet should work well here. If the Internet doesn't work well here, what hope is there for the rest of the continent and the rest of the developing world? For now, unfortunately, the Internet in Senegal—as is the case in much of sub-Saharan Africa—is subject to many political, economic, and educational obstacles that need to be overcome before it can begin to play any kind of widespread role in the country.

After well over a decade of Senegalese government policy to bring the Internet to the people, Olivier Sagna, secretary-general of OSIRIS, wrote in an editorial responding to the April 2010 study: "Finally, it is clear that the main obstacles to using information and communications technologies are the high cost of communications and computers, the poor quality of service, insufficient coverage of networks, illiteracy, computer illiteracy, and the poor availability of local content. Beyond being able to precisely measure what, up until now, have just been estimates, the goal of this study—should it be continued on a regular basis—is to clearly indicate the magnitude of work needed to be done to significantly diminish the impact of the digital divide." The editorial concludes: "It remains urgent, therefore, to give fewer speeches and more actions that fight effectively against the digital divide and to develop an information society."[5]

Amadou Top: The Boubou-Wearing Geek (1947–1995)

One person who has tried to overcome these hurdles looks nothing like a traditional Silicon Valley geek. When in Senegal, Amadou Top usually prefers to wear traditional dress, the Senegal *boubou:* a long, flowing shirt and matching loose pants, accompanied by a sleeveless and equally very loose vest on top of the shirt. The president of OSIRIS is an imposing figure—with big hands and a glowing smile. If his office weren't in a middle-class neighborhood—where many European expatriates live—he probably would fit based on appearances alone right into any traditional Senegalese village. But Top is mainly known as the president of OSIRIS, a cornerstone of Senegalese civil society that seeks to improve the availability and quality of Internet access.[6]

Since August 1999, Top has written a monthly e-mail newsletter, known by its French acronym, *Batik*, which goes out to thousands of readers both in and outside of Senegal. He has meticulously chronicled every conference, every new law, and even the price of Internet connectivity (and mobile phones). He advocates for greater transparency, competition, and level of service in his home country. He has the ear of the president of Senegal, Abdoulaye Wade; he has represented Senegal at international conferences; and for a few years he was a vice president of the Digital Solidarity Fund, a Swiss-based NGO conceived of by President Wade. He's not so much the mastermind behind Internet deployment in Senegal, but rather is someone who agitates and seems to have a discrete amount of influence.

Top is among the lucky few Senegalese who have a dedicated connection to the Internet. In a way, he stands as someone who is the exact epitome of what Senegal's leadership would like the entire country to be: determined, smart, ambitious, locally educated, and using the Internet to prosper. If Top had grown up in Europe or North America, or any other more developed country, there is no doubt that he would be an even more successful entrepreneur. In a way, his experience of living and working abroad has only frustrated him more, as he is well aware of how Senegal should be.

However, despite his, the Senegalese government's, and other foreign organizations' best efforts to change the Internet regime in Senegal, the country still has a great deal of political and economic realities that it must overcome before it can realize its potential online. The problem is that the country's leadership doesn't seem to realize this and continually attempts to drive the country forward, without realizing why so much of the country continues to drag behind. When the Internet first arrived in Senegal in the mid-1990s, it did so at a time when the state telecommunications monopoly still retained a great deal of control, which kept the price of access artificially high. Further still, Senegal's core lack of education and a comprehensive national development strategy has retarded Internet development. By contrast, when Amadou Top first saw the Internet in 1995 at the age of forty-nine, he had already had a university education and had lived and worked in Europe. He returned so he could help his country advance. In other words, he was ready. Today, Senegal as a nation, and like much of sub-Saharan Africa, cannot fully be ready until it overcomes much more basic obstacles.

In 1995, Top and the other members of a small delegation of Senegalese business leaders sponsored by USAID and the American embassy in Dakar stood amid a crowded hall at the Consumer Electronics Show in Las Vegas. They were there, along with thousands of others, to catch a demo of the recently released

Netscape Navigator 1.0. In fact, it was a bit hard to see, even for a tall man like Top—he jerked his head from side to side and shuffled around, doing his best to catch a glimpse of this browser.

As a browser window appeared projected onto a wall-sized screen, Top could not believe what he was seeing—a way to display text, images, sound, and video—together, in an interactive way. He knew then that his days of dabbling in a text-only environment were over.

At the time, Top had his own company in Dakar as a computer reseller, ATI. He mainly sold the IBM AS/400 (a piece of large-scale computer hardware) to multinational companies, the government, and universities, but he had wanted to move beyond just sales. He had thought about starting a text-only dial-up Bulletin Board System, but had been advised against it by a young American NASA engineer that he'd recently met who told him that BBSes were dead: "It's all about the Internet now!"[7] At the time, Top had no concept of what the Internet looked like, nor what it meant. It might as well have been a Chinese word.

In the United States in the early 1990s, Internet service providers (ISPs) began to pop up all over the place. These ranged from small providers to large nationwide companies, such as America Online and CompuServe, which previously had only offered access to their proprietary network. Top knew that despite the slow connection to each of these services, the fact that local calls in the United States were (and still are) sold at a low-cost flat rate meant that it was significantly easier for an American to get online than an African. In 1995, there were no local ISPs in Senegal. The only way to get online was to make expensive calls to Europe or North America. Top figured that while the Internet was a fascinating new medium, there wouldn't be much use for it in Senegal, given that it would essentially only be available to the government, universities, and large enterprises.

"We thought it would be something that would have restrictive access, at least for Africa," Top recalled.[8] Still, he thought the technology was "fascinating."

AMADOU TOP WAS BORN in Dakar in 1947, just one year after IBM came to Senegal at the invitation of the French colonial government. Initially IBM provided logistic support, computers, and software for Senegal's railroads and those of neighboring countries in French West Africa. Dakar was the administrative, economic, and political capital of the entire region, and IBM had thought that it would be profitable to serve France's colonial empire in Africa, just as it had in French Indo-China more than a decade earlier.

Top's father was from the interior of the country, near the Saloum delta, and was a former conscripted soldier in the French colonial military and, later, a park ranger. His mother bore and raised Top and his ten brothers and sisters. As a student, young Amadou Top excelled in math, but didn't have any precise focus or goal set up for himself. His parents were supportive and were pleased with his success, but didn't push him in any particular direction. Toward the end of high school he thought that he might become a teacher, as that was a well-respected profession and was something that "everyone wanted to do."[9]

After dabbling for a short time in communist political philosophy as a quasi-activist inspired by the Cultural Revolution in China, Top enrolled at the University of Dakar to study economics and philosophy in 1969. After two years, though, he learned that IBM was holding a competition for new interns. Top registered to take the three-day exam that consisted of math and logic problems. He was one of six winners from a pool of hundreds of Senegalese applicants.

He spent the first four years working for IBM in Senegal, cycling through various positions, first as an analyst, then as a programmer, and finally as a systems engineer. By 1975, he was sent to Versailles, France, for a two-year training program and worked mainly as a senior analyst—one of twelve analysts who studied the Senegalese market and how IBM could best expand in Senegal and in the region.

Top got another big break in 1982, when he was chosen as one of ten African IBM employees to travel to France and the United States to be trained on the new IBM PC. He stayed at IBM for over a decade, and after a short stint working for the Ministry of Finance, Top entered a nationwide programming contest and again won the top prize, for an accounting and inventory program that he wrote for the IBM PC.

He could have easily stayed at IBM or at the Ministry of Finance, Top says, and could have had a very comfortable and successful career there. But as he explains it, the "development virus" got hold of him, and he needed to get out into the world. He thought he could do a better job of helping Senegal by starting his own company, which he did in 1987. In a country that had just a minuscule IT sector in the 1980s, Top emerged as one of the business leaders that USAID selected to attend CES in Las Vegas.

But Top's story is not one that could have happened in many other countries in Africa—Senegal's history made it possible.

From European Colonization to the Expansion of Communications Technology (1000–1996)

Various ethnic groups and tribes have inhabited Senegal for centuries, and Islam arrived in the region by the eleventh century. When the Europeans (Portuguese) came to Senegal in the fifteenth century, the indigenous Wolof Empire was at its height. Local legend says that the name "Senegal" derives from the Wolof phrase "*sunu gaal*," which means "our canoe"—a possible misunderstanding between Portuguese sailors and the first Wolof men they encountered at sea.

Later in the century, the Portuguese king ordered a fort to be built at the mouth of the River Senegal (near present-day Saint-Louis, in the northwest corner of the country), and in 1468 he decreed a monopoly on the slave trade south of the river. By the sixteenth century the French arrived as well and began to overtake the Portuguese as the dominant European trading partner, especially when the French began to trade guns for local products, including gold and ivory, something that the Portuguese had previously refused to do.[10]

The French, in turn, faced competition and sometimes piracy and open warfare from the English and then the Dutch (the slave-trading island of Gorée was named for a delta island in South Holland, the Netherlands), as each European nation battled all across West Africa to bring back the most of these exotic goods. However, toward the end of the seventeenth century, the French had secured control of what is now northern Senegal, between Gorée Island (near Dakar) and founded Saint-Louis, a city named for Louis XIV.[11] Many of the port cities, like Saint-Louis, became major slave-trading hubs. For the next few centuries, the French firmly secured their grip on what is now modern-day Senegal and in 1857 established a military post at Ndakarou (later, "Dakar") on the peninsula opposite Gorée.

The nineteenth century saw a great consolidation of French power in Senegal, and many new communication infrastructure projects, including a telegraph line via undersea cable connecting Saint-Louis and Gorée to Spain and then France in 1862. In the early twentieth century—when French West Africa was formally organized and its capital set first in Saint-Louis, and later, Dakar—radio arrived, primarily for use by the colonial administration in communicating with its outposts inland and with ships at sea. The first "wireless telegraph" station was established in Dakar in 1928. Between the 1930s and the 1950s, French-language radio stations (mostly concentrated in Dakar and Saint-Louis) provided upper-class Senegalese and French people with a proto-Internet, a connection with the outside world. As public telephone connections became more developed in the first half of the twentieth century, users of Senegalese telephones were essentially considered "long-distance subscribers" of the colonial base in Paris.[12]

In 1960, Senegal became independent, and a peaceful and democratic transition to a new republic occurred, with Léopold Sédar Senghor as its first president. As a Senegalese-born, French-educated Catholic, Senghor represented both the educated upper class of Senegal and the general population, as his Senegalese roots made him popular at home. This democratic precedent that Senghor set paved the way for Senegal's political and economic stability throughout the twentieth century—he served nearly five terms as president, from 1960 until 1980, resigning his post before completing his fifth term. (However, in the fifty years of Senegal's independent history, there have been changes in government only twice, and only once between parties—in other words, one could argue that while Senegal is politically stable, it is not democratically competitive.)

In 1962, when the radio tax was eliminated, the number of radios shot up to 200,000 nationwide. That same year, the Senegalese government expanded radio to the rural interior by creating 145 "listening posts" scattered around the country, so that teachers, nurses, and other leaders in those areas could act as conduits to transmit important information to the masses. As the Senegalese professor and a founding member of OSIRIS Olivier Sagna writes: "These were supervised by teams of volunteers, including teachers, nurses, rural coordinators, or war veterans, and were regarded as tools for the stimulation and education of the population. This system could be considered the predecessor of the community telecenters, interspersed to provide Internet."[13]

For all of the ills of French colonialism, they did leave behind an entire infrastructure of education, transportation, and communications that did not previously exist. Traditionally, as had been the case in North America and Europe, the PTT (Post, Telephone, and Telegraph) office performed all communications functions, and was a state monopoly. Senegal's PTT functioned normally for twenty-five years after independence, but later was subject to what telecommunications experts call "bifurcation," or the policy act of separating postal services and telecommunications services. This was an important initial step in reducing prices through demonopolization. (In Africa, Ethiopia was the first to bifurcate its telecom company in 1967.) In 1985, Sonatel was carved out of Senegal's traditional PTT, leaving the Post to perform mail delivery functions, while Sonatel continued telecommunications services.

Unlike in the United States and Europe, which began to demonopolize and allow competition in the telecom sector in the 1980s, demonopolization largely did not take place in Africa until later. The continued monopoly meant that prices for telephone calls remained (and to some degree still remain) artificially high. Further, for many years it was difficult to even get an actual landline installed in one's home or workplace, as many telecom monopolies were overwhelmed with demand, corruption, and incompetence. Beyond the bureaucratic delays, the actual infrastructure was of inferior quality and was not properly maintained. In the mid-1980s, 60 percent of all Senegalese telephone lines were unusable during the rainy season (June to September), and 10 to 15 percent were unusable during the rest of the year.[14] With zero competition, consumers had no alternative and the monopolies had no incentive to change.

All across the continent, it was common in the 1980s and 1990s for people and businesses to wait years, sometimes even as long as a decade, before receiving a telephone line. In fact, according to a 1998 study of twenty-seven African nations by the International Telecommunications Union, between 1990 and 1996, waiting time to get a phone line installed improved in eleven of the twenty-seven countries, but in the remaining sixteen that time actually increased. Further, in twelve of those sixteen the waiting time actually doubled. In Senegal, this meant that there were three lines for every one thousand people—two-thirds of which were in the Dakar area.

During the 1980s, Sonatel was a disastrously managed company. A 1985 internal company report "showed that only 30% of the bills were paid within the legal delay of 30 days, while 55% were paid within 60 days" and that "a survey performed by an independent audit firm showed that 30% of the sums reclaimed from clients had already been paid, while a lot of unpaid ones were never reclaimed. However, the same report asserted that the second cause of the large amount of unpaid bills was 'the tolerance granted to some customers, due in part to their network of influence.'"[15]

While Senegal's floundering continued, elsewhere in the world very basic Internet access was starting to take root. In the late 1980s, Randy Bush, an American software and network engineer, helped to found the Network Startup Resource Center, a National Science Foundation-funded operation that aimed to get various developing countries online. The NSRC brought various countries in southern Africa (including Botswana, South Africa, Namibia, and others) online starting in 1988, and by the early 1990s had moved on to Peru, Egypt, and Guinea.

In 1992, Sonatel had the problem of needing to expand its scope of service while being unable to hire more staff and expand its bank of public phones. As a solution, the state-run company decided to pilot test a network of public phone booths, or "telecenters," in four sites across the country.[16] Telecenters are essentially small, privately owned rooms, sometimes part of a home, that have publicly accessible phones for people to call to and from. Customers come and pay for the time that they use. In countries where most people don't have landlines, the telecenters serve as an easy and cheap way to make calls. Further, they create an entire new sector of small businesses.

For the first two years of the telecenter program, Sonatel imposed minimum distance requirements between telecenters, as a way to keep the competition controlled. That rule was later relaxed, and the number of telecenters continued to explode, with more than half of them being in the capital, Dakar. By 1998, there were nearly seven thousand spread across the country, and the typical per-unit cost for customers making calls at telecenters dropped from 100 CFA francs ($0.20) to 65 CFA francs ($0.12), cutting the profit margin to just five CFA francs ($0.01) on each unit called. By 2000, the number of Senegalese telecenters grew to almost ten thousand—the largest single number for any African country.[17]

While Sonatel was starting to dabble into its private franchise model, the country was also undergoing major financial reforms. In 1948, France created a special currency for its African colonies, known as the CFA franc (the words behind the French acronym have changed over time—currently, in West

Africa, they mean "African Financial Community").[18] The CFA franc, which has since been adopted by a total of thirteen countries (not all of which are former French colonies), was pegged to the French franc beginning on the day that France ratified the Bretton Woods Agreement, on December 26, 1948. In 1958, when the French franc was devalued, the CFA-to-French franc exchange continued with one hundred CFA francs being worth two new French francs.

However, as Senegal and the other former French colonies became independent, France continued to prop up the value of the CFA franc, and as such, the cost of goods exported from these African countries remained artificially high as well. In January 1994, France decided that it would no longer support the CFA franc, and declared that it would only make future aid payments to CFA countries if they submitted to the monetary and economic will of the International Monetary Fund (IMF), which had recommended the devaluation as a way to boost exports. Further, if those countries agreed to devalue their currency, then they would be eligible, for the first time, to directly receive IMF loans. While the devaluation helped CFA-nation exports, they also doubled the price of every imported product overnight. Protests, labor disputes, and demonstrations of public disapproval flared up in many cities.[19]

One of the IMF's recommendations called for privatizing all state-owned companies, including Sonatel, as a way to create more competition and more efficiency. By February 1995, Sonatel's board of directors created a plan to partially privatize the company. The following year, the company was auctioned off, and France Telecom ended up with 33.33 percent of Sonatel, which later grew to 42.33 percent, making it the largest single shareholder.[20] Few Senegalese citizens or businesses were consulted prior to the sale. Worse still, the deal preserved Sonatel's monopoly on fixed line and international calls until 2004.

At the end of the monopoly period in January 2004, Amadou Top wrote as the headline of his monthly *Batik* e-mail newsletter: "Finally ! ! !" Top lamented the fact that the government had not allowed Sonatel to be taken over by a Senegalese company, or at least a pan-African consortium:

> Instead, Sonatel was transformed into a private company with Senegalese ownership, but with capital held mainly by a foreign entity— France Telecom. Moreover, by granting [it] monopoly rights with regards [to] fixed and international telephony, the Senegalese state has for seven years delivered to a foreign entity, rights on matters that it prohibits its own citizens. It is as if we had returned to colonial times.[21]

Despite the monopoly's ending in 2004, an ISP to compete with Sonatel did not start up until January 2009. This delay occurred despite the fact that the government approved a third license for $200 million in September 2007 to the

Sudan-based Sudatel.[22] Expresso's 2009 launch finally broke the long-held Sonatel monopoly on Internet service.

IN THE YEARS AFTER THE 1996 privatization of Sonatel, public Internet access started to become available in Senegal, with the presidential palace being the first place to be wired. Sonatel began selling Internet access—which it received from a satellite connection—to the public in April 1996.[23]

On July 3, 1996, Métissacana opened its doors in downtown Dakar, claiming to be the first cybercafé in all of West Africa, and the first on the continent outside South Africa. Founded by Michel Mavros, a French national but longtime resident of Senegal, Oumou Sy, his Senegalese wife and a fashion designer, and Alexis Sikorsky, a Swiss computer technician, Métissacana was open twenty-four hours a day, 365 days a year. The cybercafé charged 1,500 CFA francs (about $3) for one hour of service, or less with a monthly or annual subscription. In addition to fourteen computer terminals, Métissacana also provided telephone, fax, and office services. Still, Métissacana couldn't have had that many customers in its first year—on October 30, 1996, Mohamet Diop, the data network manager for Sonatel, estimated that there were six hundred Internet users in the entire country. Métissacana closed as a cybercafé and as a residential ISP in July 2002 in order to protest what it called Sonatel's "abuse of monopoly."[24]

THROUGHOUT THE LATE 1990s and early 2000s, cybercafés began to spread across Dakar and the rest of the country. In fact, many telecenters expanded their traditional landline-only services to include computers as well, sometimes even reducing the number of telephones they had in order to make room for computers. As more and more cybercafés opened, the hourly rate for Internet access continued to fall dramatically, said Blaise Rodriguez. This Senegalese cybercafé owner founded a combined computer school and cybercafé attached to his home in 1999, near the American-run Baobab Center in SICAP Baobab, a middle-class neighborhood in Dakar. In an interview, he said:

> I should say that the raising and lowering of prices had to do with the cybercafé in Sacré-Coeur. They were playing with the prices. After he opened at 1500 CFA [around $3], and then dropped to 1000 CFA [around $2] and finally to 500 CFA [around $1], all the customers went to him. We were forced to drop to 500 CFA as well. And after the end of 2001, as that was coinciding with Senegal's qualification for the World Cup, he dropped further to 250 CFA. We all had to follow him and to set our prices at 300 CFA. So you can say that since 2000, cybercafés in Dakar have all been between 500 and 300 CFA. The most expensive will charge 500 CFA, and the least expensive at 300 CFA—but there aren't any more 1,000 CFA cybercafés in Dakar.[25]

Unfortunately, cybercafés open and close too often to have any reliable statistics on the number of them anywhere in Senegal—but these days, at least in the capital, it feels like they're everywhere.

But 1996 was not only a big year for Internet access in Senegal, with the beginning of its first cybercafé, but in other parts of the continent as well. That same year, the United States began its first and most ambitious policy of getting other African countries online. The Leland Initiative was a $15 million, twenty-one-country, five-year program to help various nations across Africa actually get online. Senegal was among the chosen few.

Leland, named for the late Texas congressman Mickey Leland, was designed to make sure that these sub-Saharan countries did not fall further behind on their path to greater economic development. As USAID states on its website: "Africa needs access to the powerful information and communication tools of the Internet in order to obtain the resources and efficiency essential for sustainable development."[26]

Leland had three major objectives: first, to "create an enabling policy environment." That meant that the costs should be kept down through private sector competition as a way to maximize Internet users. Second, to "create a sustainable supply of Internet services," which meant expanding the number of local ISPs and making sure that there is access even in rural areas, and also having other members of civil society actively participate in the discussions and observance of Internet development in the country. Third, to "enhance Internet use for sustainable development," essentially meaning that the Internet would aid health, environment, government, education, and other related areas.

In order to become a Leland country, the in-country USAID office had to send a letter to USAID headquarters in Washington, D.C., saying that Internet assistance was both "desirable and feasible." Then, there would be more paperwork, including a Memorandum of Understanding, and finally a Plan of Action. Leland was a bold step forward toward assuring that African countries would be able to have an initial and growing Internet sector. Leland brought the first Internet connections to Mali, Madagascar, Mozambique, and Guinea in 1996 and also got Rwanda, Ivory Coast, and Benin online in 1997.[27]

Amadou Top Introduces the Government to the Internet, and in Turn, to the Country (1995–1999)

After Amadou Top returned from the CES conference in January 1995, he decided to found a subsidiary of his company, ATI, and called it "Interactive," as a way of promoting Internet access in Senegal. He owned half of Interactive, and ATI owned the other half. Top then set out to create a collection of webpages, with video, audio, and images, as a way to demonstrate to local businesses what exactly the Internet could do for them. As Senegal did not have an Internet connection at the time, the webpages were designed to run on a "localhost" computer—that is, a computer without a connection to the outside Internet. They were meant to demonstrate how information could be organized, presented, and then retrieved online. Top wanted the presentation to have the same jaw-dropping effect on others that it had had on him the first time he saw it. His first site was a sample for PANA, the Pan-African News Agency, complete with news stories, photographs, and even video. He spent weeks tweaking it, making sure that every image was placed properly, that every sound and movie file would work just as it was supposed to.

By the end of 1995, Top started making his presentation to various high-level businesses that were set up in Dakar, including one to a meeting of international bankers and other government finance officials. He had a computer set up, which was then connected to a projector, much in the same way as he had seen the Internet first demonstrated to him. However, unknown to him at the time, one of Abdoulaye Wade's staff was in attendance at the banker meeting. Wade, the current president of Senegal (until 2012), was at the time a Minister of State. This staffer alerted Wade's office to this presentation about the Internet, and Wade promptly requested that Top come to his office for an encore performance.

Top wasn't nervous at all—after all, this was a presentation he had given many times previously. Accompanied by three of his staff, Top set up the computer as normal, in Wade's office. While Wade wasn't a geek by any means, he did have some basic knowledge of computers. Most Senegalese government ministers had computers in their offices by that point, anyway. Top launched the PANA site and showed Wade the photos, the video, and the homepage. Wade remained fixated on the screen, his eyes darting between browser windows and Top, who stood in front of the room, again, wearing a traditional boubou. Wade moved up in his chair and asked a lot of questions: "Are Windows PC machines enough? How can we use this in the countryside? Will our students have access to new libraries?"

Top answered every question that Wade had, and emphasized the interactivity that the Internet created, emphasized the simplicity of bringing the Internet to Senegal—all the country needed was access and bandwidth. "The day that this comes, it will be a revolution," Top told Wade.[28]

The meeting took nearly two hours, nearly twice as long as usual, and by the end Wade was completely enraptured.

"[Wade was] the first important political figure to get that there was something that was going to change," Top said.

Less than a week later, Top found himself giving the same presentation to the chief of staff of President Abdou Diouf—Top is certain that it was Wade who made that meeting happen. Later that year, the first Internet connection arrived in Senegal, a satellite connection of 64 kbps for the whole country.

At the time, many online Americans had 33.6 kbps modems on their desks—or half the bandwidth of the entire country of Senegal.

OVER THE NEXT TWO YEARS, as the Internet slowly started to seep through Senegalese society, Amadou Top decided to create a more formalized structure to promote Internet access. This organization, OSIRIS, lobbied the government to spend more public money on Internet access—just like his demonstrations that he had done before. However, this time, with leaders from the public, private, and academic sectors, OSIRIS could be an effective organization that would try to understand how to make the Internet work better in their home country.

In the early and mid-1990s, a similar organization, the Internet Society (ISOC) was organizing local chapters in countries on a global scale as a way to "promote the open development, evolution, and use of the Internet for the benefit of all people throughout the world," as it states on its website. At the time of ISOC's founding, the Internet remained largely an academic endeavor, and there were not many ordinary people who had heard of it, nor was there a high level of commercial Internet service available in Senegal or in most countries in the world.

Beginning in 1993, Randy Bush of the NRSC, and two original members of ISOC, George Sadowsky and Larry Landweber, began to set up special workshops around the same time as that year's INET conference, ISOC's annual meeting. The workshops were for attendees from developing countries who wanted to learn more about how, technically speaking, to bring the Internet to their respective countries. These workshops continued throughout the 1990s, and as Landweber later put it: "We essentially produced a cadre of people who knew about the Net in a number of other countries."[29] These initial workshops, combined with the efforts of the National Startup Resource Center, later morphed into the Leland Initiative.

IN JANUARY 1998, USAID wrote up its findings for the "Leland Initiative Report, Telecommunications Sector Assessment: Senegal," the American assistance plan for the Senegalese telecommunications industry. Even though by 1998 there were relatively few Internet users, USAID was rather impressed with what it saw in Senegal. In fact, the opening sentence of the report states: "In comparison with its neighbors, Senegal possesses ones of the most promising telecommunications sectors in West Africa."[30]

The report noted the important initial steps that Senegal had taken to make its Internet infrastructure work better. For example, Sonatel had changed to an all-digital switching network, had near-complete digitization of its transmission infrastructure, and had privatized its national operator as a way of making it more efficient and more effective. The report also pointed out that Sonatel was "in the process" of establishing high-speed fiber links between

Dakar and the regional capitals. A 1997 diplomatic cable from the American foreign communications officer in Senegal described the country to the Department of Commerce in Washington, D.C., as "now well on its way to providing Africa's first fully digital telephone network."[31] On top of it all, within less than two years of bringing the country online, Sonatel increased the level of international bandwidth in early 1998 from 64 kbps to 384 kbps.

However, while the report noted Senegal's initial steps and potential for greater success, it outlined four major points that would "accelerate and facilitate the transition of Senegal to an Information Society."

These four points were: (a) "rapid establishment of an effective, independent regulatory commission," (b) "substantial strengthening in formulating strategic telecommunications sector reform policies," (c) "strengthened capacity for encouraging and managing increased competition in the telecom sector," and finally (d) "development of a consistent national policy on the transition to an information society."[32]

Essentially, USAID was suggesting that Senegal create its own version of the Federal Communications Commission (FCC), which could act as a regulatory body to ensure proper levels of competition in the telecom sector and to make sure that one company—like the incumbent monopoly, Sonatel—would not totally dominate its new competitors. The report also recommended a new set of laws that should set a new vision for an "information society," and encompassed all kinds of aspects, ranging from how telecom companies compete with one another to how the government deals with various future technology issues, such as cyber crime and the use of digital cryptography.

While the government may have dragged its feet in getting a broad telecommunications "vision" in place, by 1998 Amadou Top already had a vision of his own. He created OSIRIS, a French acronym for the "Observatory on Information Systems, Networks and Information Superhighways in Senegal," with a nod to the ancient Egyptian god of the afterlife. Further, OSIRIS was probably the first such Internet-related civil society organization anywhere in sub-Saharan Africa.

On one Saturday morning in March 1998, Top convened a meeting of no more than six or seven of his tech-inclined colleagues in his office. He had intended for OSIRIS to be small, so it would not get bogged down in bureaucratic excess. This organization, he knew, would be streamlined and would create a clear means for him and his most trusted colleagues to shape tech policy. Because OSIRIS organized so early in the history of the development of Senegal, it was able to have a disproportionate influence on government policy.

Only a few weeks after the official launch of OSIRIS, Amadou Top had another significant role to play in deploying the Internet in Senegal. On May 25, 1998, Senegal conducted elections for all 140 of the country's parliamentary seats. In the run-up to the balloting, as usual, the campaign was hard-fought.

Abdoulaye Wade, who by this time had left the government and was the head of the Senegalese Democratic Party (PDS), rival to the incumbent Socialist Party (PS), alleged charges of fraud by that party. As a way to allay those fears and put down similar allegations, President Abdou Diouf nominated army general Lamine Cissé as Minister of the Interior in 1997. One of General Cissé's jobs was to provide security at the polls during the election, and to make sure that the electoral rolls were accurate.[33]

Within a couple of weeks of the election, General Cissé had to ensure that the more than 3 million voter cards were distributed, and that there were accurate lists of every eligible voter. With a lot of tension around the elections rolls, not all Senegalese people—least of all, the political opposition—believed that these rolls would be completely accurate. General Cissé did not think that there was enough time to check all of them, nor to get the necessary lists out to the country in time. So, he called Amadou Top to see if there wasn't a way to get the electorate to trust the election rolls.

"The only way to do it in the little time we have left is to put them online," Top replied.[34]

So Top, with the help of his company, ATI, took the government's electoral roll and posted it on an official government website within a few days. A United Nations report later called this decision "a major factor in organizing the poll, which was contested from the outset but all parties involved approved the results."[35]

While Top says that this was the first time that any country in the world put its electoral roll on the Internet, there were some problems with putting so much personal data online so quickly. Top and his team not only included people's names, but their birth dates, addresses, and mothers' names as alternative identifying data. Some Senegalese citizens who were living abroad as illegal immigrants found themselves on this list, and were fearful that they would be found out in their new countries based on this data. Many quickly made calls to the Ministry of the Interior to get the addresses taken down.

Other citizens hounded the government for having put their mothers' names on the website. They were afraid that listing their mothers' name could leave them susceptible to "curses" laid upon their family. More practically, many illiterate people also used their mother's name as a password for their bank card. Top quickly took down the list of mothers' names from the registry as well.

In the end, while there may have been some hiccups along the way, Senegal became the first country in the world to make such a bold step of putting a country's election rolls online.

BY END OF THE SUMMER of the following year, August 1999, Amadou Top had not only created OSIRIS, but he had also launched the organization's website, and put out the first e-mail and online monthly newsletter, *Batik*.

The newsletter is a French acronym for the "Analysis Bulletin on Information and Communications Technologies," likely the first locally produced publication on the continent devoted to following the machinations, ins-and-outs, and day-to-day goings-on of Internet life in Africa. Each monthly edition of Batik begins with an editorial by Amadou Top, in which he usually gives his opinion on that month's technological news. The first edition of *Batik* began this way:

> Everyone talks about changes that are going to engender the ways of learning and of doing [and] that are going to accompany all these technologies that come together as part of the Internet. We would like, for our part, to go beyond these technical discussions, and to observe in a vigilant manner, the way that a society like ours realistically behaves when it comes into contact with new information and communications technologies, and especially how it appropriates them on a daily basis, with its disadvantages and its advantages, to its own rhythm and through its own cultural prisms.[36]

The first news item that month also noted that at that year's INET '99 conference, held in San Jose, California, several Senegalese had participated in the pre-conference workshops, including Mouhamed Tidiane Seck, who as a computer science professor was one of Senegal's very first Internet users in 1990. (He later became the Minister of Technology in 2003.)[37]

Again, the fact that Top has been a longtime agitator for the Senegalese IT industry makes him unique, particularly when compared to other sub-Saharan African countries. Without his efforts of chronicling and lobbying, Senegal certainly would be much further behind than it is today.

Failed American Influence and Failed Senegalese Regulation (2000–2002)

By 2000, there was more than Amadou Top and just USAID that wanted to help Senegal get more people online. Just as for many decades (and even centuries) in the past, there were developed countries that considered it their mission to try to come to Africa and help it develop. In this case, helping Senegal get online. After all, the logic goes, Senegal "missed" the industrial revolution, and so it had better not miss the information revolution.

USAID had a clear plan, a mission to create a foundation of sorts for the Internet in Senegal and let it develop on its own organically. The early 2000s also had a series of designed institutions and projects, some of which originated from inside Senegal, and others, from outside the country. Regardless of their origins, the collective failures of these projects illustrates that in a country like Senegal, basic domestic factors like education and bureaucratic competence may not be as developed as these outside actors would like them to be.

The first major Internet project brought to Senegal was Joko, a joint American and Senegalese venture. Initially, Senegal's best-known musician and Goodwill Ambassador of the United Nations International Children's Education Fund (UNICEF), Youssou N'Dour, kicked off the Senegalese side of things. N'Dour, who has tried to help his home country in numerous ways as he has risen to international stardom, has particularly focused on Medina, the poor neighborhood in Dakar where he grew up. In 2000, he had the idea creating a network of cyber centers, or "Joko clubs" (*joko* being a Wolof word for connection, or link), which would be used as conduits so that African children could communicate and engage in cultural exchanges between themselves and other kids around the world.[38]

N'Dour ended up partnering with Hewlett-Packard, which at the time was beginning its own "e-Inclusion" program as a way to "address the needs of the poor and future growth of the company," according to Janine Firpo, the then-manager of external collaboration for HP's e-Inclusion initiative.[39]

HP budgeted millions of dollars to set up the Joko clubs, which essentially were glorified cybercafés, with a little bit of training and cultural exchange tacked on. At the January 2001 World Economic Forum in Davos, Switzerland, N'Dour himself touted the new Joko clubs. The first one opened in August 2001, and by 2002 HP and N'Dour's new company, YND, had set up four in the entire country. The project had an initial budget of $3 million, an extraordinary amount of money in sub-Saharan Africa.[40]

The project received a lot of attention in the media as it progressed, and HP touted its own success. In a July 2002 report, Firpo and one of the cofounders, Lisa Carney (now Lisa Goldman) wrote:

> The response was overwhelming, and over 1,500 people were on the waiting list by the time this first session ended. At the two pilot Joko Clubs, prices for the initiation courses were set at 3000 FCFA [$6] initiation fee plus 3000 FCFA for a month of training. The Medina Joko Club reports teaching more than 1,000 new students, including 80 illiterates. Ngoundiane taught more than 1,200 students, 70 analphabetes among them. Ngoundiane reports that so many other rural communities have sent delegations to spend a week getting initial computer training that they are considering opening a "bed and breakfast" to accommodate them. These villagers say they feel at ease in a familiar, rural setting—so unlike the urban environment in Dakar.[41]

However, it may have gotten a little too successful in these particular locations. One article in the *Boston Globe* published in October 2002 found that the in-country managers abused the corporate generosity of HP:

> [Amadou Kromma, a Joko club manager], for example, [preaches] the doctrine of universal access at low cost, sucked in money from what he

seemed to think was a bottomless pit. He charged basically nothing for courses and offered jobs to many of his friends from around the Medina in traditional African custom: As soon as anyone has a little bit of money, they are often expected to simply spread the wealth.

The center now has about 20 employees, consisting in large part of people sitting on the front porch without much to do. After receiving about 30 computers from Hewlett-Packard, the club requested dozens more, along with a few four-wheel-drive vehicles and five motorcycles.

"There is a belief in Africa that an American is wealthy beyond belief and they should be giving money because they have more than they could possibly deal with," said Janine Firpo of Hewlett-Packard's E-Inclusion project who serves on Joko's board. Firpo added, however, that there's a certain logic behind that point. After all, Hewlett-Packard, despite its recent woes, is worth 10 times as much, in economic valuation, as Senegal.[42]

The *Globe* noted that at the time of the article there were four Joko clubs. Despite the fact that the Joko website claimed that there would be fifty more by the end of the year, the entire project shut down before the end of 2002.[43]

In addition to the project's incompetence and nepotism, the planners of Joko realized that they could not compete with profit-oriented cybercafés and telecenters-turned-cybercafés that were spreading out all over the country at the same time.

"In order for them to continue to be maintained, [that] would [have required] ongoing investment, grants, or philanthropic funding," Firpo said.[44]

Clearly, despite its noble intentions, HP was not prepared to write those checks.

WHILE HP AND N'DOUR'S JOKO project didn't really make it off the ground, at the end of 2001 the government of Senegal clearly had the intention of pushing its tech industry forward. The president signed into law the new thirty-two-page comprehensive telecommunications bill (Law No. 2001–15 of December 27, 2001). The law defines a "national strategy" based on:

- the immediate liberalization of value-added services in the context of a free market
- the 1997 privatization of Sonatel, which allowed it to become more competitive
- the opportunity for private companies to establish local cellular networks
- the effectiveness of universal access across the country, notably in rural areas and at affordable prices
- short-term liberalization of the fixed-line and international access in 2004[45]

The language of the 2001 telecommunications law almost exactly mirrors what USAID advised Senegal to undertake three years earlier, calling for an "effective independent regulatory commission" and "a consistent national policy on the transition to an information society." The law goes on to call for an "efficient and transparent judicial framework," with the aim of creating a free and fair market "for the benefit of [all] markets." The text of the law includes a lot of free market rhetoric, saying that the law will conform to international standards and acknowledging that it will fight against "the abuse of power."

One of the most important features of the 2001 law is the creation of the Agence de Régulation des Télécommunications, or ART (which in 2006 came to be known as the ARTP, or the Agence de Régulation des Télécommunications et des Postes, as its mandate was expanded). Established in March 2002, this new body would be the executive, independent authority to enforce new telecom laws and policies. The ARTP's job is to advise the president, provide consultations on relevant national legislation, and represent the Republic of Senegal in international institutions such as the International Telecommunications Union. The ARTP also acts as a body for arbitration between citizens, providers, consumer groups, and the state itself.

The ARTP is mainly charged with maintaining competition between mobile phone companies. When it was created in 2002, there were only two mobile operators, Sonatel (operating under the brand "Orange") and Sentel ("Tigo"), and so if one of these companies were to act in an improper way, it would be the job of the ARTP to fine the company. In order to expand mobile, fixed-line telephone, and Internet services, the ARTP is also responsible for orchestrating the breakup of Sonatel, the incumbent national telephone operator and the largest mobile and Internet provider. At the time, Sonatel controlled the only international access point via an undersea cable (inaugurated in May 2000), the Atlantis-2, which linked Europe to South America by passing through Cape Verde and Dakar.[46] In fact, Dakar is the only landing point on the African mainland for the 12,000-kilometer-long Atlantis-2 cable. When the ARTP came online, all small ISPs, including Métissacana, other cybercafés, and all the telecenters were forced to buy access directly from Sonatel.

In its first year, the ARTP (then the ART) oversaw the implementation of the SAT-3 undersea fiber-optic data cable, formally known as the South Atlantic-3/West Africa Submarine Cable (SAT-3/WASC). But, no company or country owns the cable. Instead, a consortium of thirty-six international shareholders controls it. France Telecom, the parent company of Sonatel, has one of the largest shares in the cable—15 percent, or $96 million. While the presence of the cable has led to the rapid decrease in the cost of Internet and telephone access from Senegal and other African nations, it only has further entrenched the monopoly position of Sonatel, as the company is the only Senegalese member of the consortium, through its parent company. Moreover, there was no debate, public forum, nor period of inquiry by the ARTP or anyone else to determine how the presence of SAT-3 would further ensure the monopoly. Indeed, all other Senegalese ISPs, cybercafés, and telecenters—as was the case previously, during the Atlantis-2 cable era—must still buy all of their access from Sonatel.[47]

The ARTP was also absent from the discussion surrounding the creation of the SAT-3 cable (it was a newly created organization). And it did not (and still does not) have a clear long-term policy agenda. It seems to operate haphazardly, and for the first few years of its existence, did little in the way of regulating, setting policy, or providing market studies. In fact, its own numbers show that the incumbent operator, Sonatel (now known as Orange), has 70 percent of the mobile phone market in Senegal and, given the fact that Sonatel controls the sole international Internet connection, virtually all of the Internet access. The ARTP's mismanagement is exacerbated by the fact that the ARTP has had three different directors since its inception in 2002, and each new replacement was named for what one IT analyst in Senegal, Ben Akoh, called "unclear political reasons."[48]

Ousmane Mbaye, the ARTP's former head of IT and ICT from 2002 to 2004, said in an interview in January 2007 that he left the organization partly because he didn't trust the leadership's competency.

"ART was supposed to regulate the sector and validate rates and stuff—I don't know what they're doing [now]," he said. "It's been about five years since ART was created. They're supposed to validate all the pricing [for mobile phone service and Internet service providers]. Right now they don't have the resources nor the expertise to validate pricing."[49]

But incompetence may have been more than just a lack of vision. The year after Mbaye left the ART, the director, Malick Gueye, was fired amid accusations of embezzlement. In 2005, investigators discovered a 550 million CFA franc (around $1 million) "hole" that was unaccounted for in the ART's bank records and accounts. Investigators alleged that Gueye approved exorbitantly high salaries for his staff and took an extra $200,000 for himself.[50] Gueye denied all the charges.

Promotion of the Internet in Senegal (2001–2003)

While many things were happening behind the scenes in Senegal—establishment of the 2001 telecom code and the creation of the ART—Amadou Top knew that the only way that the Internet would really take off in Senegal was if the people were able to experience it for themselves. That way, he figured, they could have the same sense of wonderment that he had experienced back in Las Vegas. From small villages in Tambacounda to the beach towns of Casamance, the Senegalese people would finally be able to explore something that only a few thousand (out of 10 million people) had known in the country up until that point. Articles in the newspaper, or even programs on television that described the Internet, were not enough. Top knew that the only way to spark people's interest was to bring the Internet to them—and thus began the "Multimedia Caravan."

Top imagined that a few vans and trucks could travel around the country, equipped with a phalanx of computers. The caravan, in turn, would use a satellite Internet connection to show ordinary people what exactly the Internet was and how they could use it. At the time, while there were some cybercafés and other Internet-equipped telecenters in Dakar, much of the rural interior of the country had never seen the Internet before. So, as a way to realize this vision, Top wrote a request on behalf of OSIRIS to the Dakar office of USAID, seeking "logistical support and procurement assistance" to send a small number of computers, a projector, a truck to serve as a mini-office and cybercafé, a minibus to transport the staff, and a van to carry the logistical materials (tent, chairs, etc.). OSIRIS also sought (and received) financial and logistical support from WorldSpace, a satellite radio company conceived of by the Ethiopian-born Noah Samara. WorldSpace custom-outfitted a truck at a cost of about $400,000, including a computer and a satellite Internet connection that would provide constant connectivity.[51]

On August 1, 2001, USAID sent Top a letter saying that it had approved the allocation of nearly $34,000 from the Leland Initiative—approximately 25 percent of the total cost.[52] WorldSpace, the Senegalese government, the Senegalese postal system, Sonatel, and others agreed to fund the remaining costs.

SO WHAT EXACTLY did the Multimedia Caravan do? Its proposal said that it would "ensure connection to internet for the visited populations and benefiting organizations," and would "ensure updated knowledge through continuous training and exchange on experiences and ideas through [digital] radio via satellite, telemedicine, tele-teaching, self-education CD-ROM, movies. . . ."[53]

The caravan started off in Thiès, a regional capital about an hour northeast of Dakar. Once it arrived there on August 11, 2001, Top and the nearly thirty others who were part of the team immediately went to work setting up an eight-computer cybercafé and projecting a computer screen for the assembled spectators. Various applications of the Internet were shown to the public, including e-mail and the creation of local, proto-online yellow pages, known by their French acronym as SIUPs.[54] A joint Senegalese-American NGO, CRESP, spearheaded these directories, and was one of the many participants of the Multimedia Caravan.

Mamadou Gaye, a codirector of CRESP, said there were many surprising results during the caravan. One of them was the time that a farmer took on an unexpected role. While on the road one day, Gaye was giving a workshop in a rural community. As usual, he asked those who would not be taking part in the activity to leave. But the people who usually left—the drivers, the guards, and those without any specific technical know-how—remained.

"We're the teachers now," one said.

"What do you mean, you're the teachers?" he asked.

"We've been trained, and now we'll teach others," another answered.[55]

Gaye was astonished that these groups, whom no one had thought to formally teach, had actually been paying attention during all of the lessons. Initially, the caravan's organizers had erroneously believed that the Internet was not for everyone and that it was only for those who had actively come to seek it out.

As Gaye said in a later interview:

> So by the end of the tour, everyone knew how to use the Internet. There was one of our guards who said: "When I'm not doing this, I'm a farmer, I raise chickens and things. Here's what I discovered on the Internet— while surfing, I found out a way to increase my crop yield, and I know how to do that now." So it's a question of information. He was just a guard. We needed him to move the computers from one place to another, and at night he didn't do anything, but he used his free time to learn about Internet research, to be trained and introduced to the Internet. You don't need university degrees to know how to use a computer.[56]

By the time the Multimedia Caravan was complete after six months, it had visited all the regions of Senegal, traveling thousands of kilometers, ending up in Nouakchott, the capital city of Mauritania, Senegal's neighbor to the north.

However, even though the Multimedia Caravan may have been an immediate success, and may have introduced hundreds, even thousands, of Senegalese people to the Internet, in the end, just like Joko, it was but a flash in the pan. At a monthly cost of around $24,000, there was no way that the Multimedia Caravan could be sustained over the long term, with the transportation of so much equipment and personnel (around twelve), plus food and lodging for all of them.[57]

Further, USAID's contribution of nearly $34,000 was contingent upon OSIRIS and the other organizations behind the Multimedia Caravan surveying Internet users and creating a database to:

> cover among others, internet users' attitudes and a listing of various services that can be provided. In addition, a monitoring plan will be developed and a qualitative evaluation will be conducted to determine if goods and services offered by the benefiting organizations have been delivered and/or performed as expected. The database will also include a directory of information in each locality visited (geography, decentralized

administration, cultural life, tourism, sites and monuments, economic productivity, etc.).[58]

Not surprisingly, this part of the project was never completed.

THE FOLLOWING YEAR, 2003, saw the beginning of the largest, most ambitious, and largest failure of Senegalese Internet promotion: the Digital Freedom Initiative, a three-year project organized by USAID.

On March 4, 2003, Ari Fleischer, the White House spokesperson, announced the DFI to the American public with a brief mention in a White House press briefing.

> Later this afternoon, the President will meet the leaders of the digital freedom initiative. This is a program to promote economic growth by transferring the benefits of information communication technology to entrepreneurs and small businesses in the developing world. Under this program, 100 volunteers will be mobilized to be sent to Senegal as the first country to benefit from this program, designed to promote growth.[59]

What Fleischer didn't mention was that this program was an extension of the groundwork that had already been laid via the Leland Initiative. Later that day, the USAID administrator, Andrew S. Natsios, gave a statement linking Leland to the DFI, and explained why Senegal was chosen as the pilot country, and what the DFI could achieve:

> Already, Senegal enjoys one of the strongest domestic information back-bones in the developing world, with cybercafés and telecenters located throughout the country. And just last May President Wade inaugurated a transatlantic cable system that connected the country to the world-wide network of high-speed fiber optic cable.
>
> Second, the DFI will facilitate the development of ICT applications that will enable small and medium-sized businesses to become more profitable, find new markets, and access credit and other inputs more easily. Over the life of the pilot activity, we envision more that than 350,000 small businesses will be involved.
>
> We plan to bring together small businesses, software firms, cyber-cafés, and volunteer ICT experts from the U.S. Together, they will identify problems and find ICT solutions that will help these businesses grow.[60]

The main premise of DFI was to achieve three objectives:

- Enable innovation through volunteer-led business and entrepreneur assistance
- Drive pro-growth legal and regulatory reform

- Leverage existing information and communications infrastructure to promote economic growth[61]

According to interviews with former DFI leaders, and DFI planning documents, the innovation portion of the project basically meant that there would be workshops and training as a way to improve the business and economic environment through the increased use of computer and Internet technology.[62] Locally, members of USAID, the U.S. embassy, their Senegalese partners, and Senegalese representatives ran the steering committee for the government. The committee met quarterly to discuss the operations of DFI Senegal and to evaluate its successes and failures.

One of the original projects that DFI created was CyberLouma, or "cyber marketplace." Opened in May 2004, CyberLouma was originally envisioned as a business-oriented cybercafé where merchants could improve their business skills. DFI's planners back in Washington, D.C., thought that it would be smart to make CyberLouma convenient to local business owners, so they decided to place it at the foot of Marché Sandaga, Dakar's biggest and busiest daily market. They thought the convenient location would make it easy for business owners to come and learn better accounting practices, and how to use spreadsheets as a way to improve their businesses.

Geekcorps, an American-run NGO, ran the day-to-day operations of Cyber-Louma. Geekcorps' premise is analogous to the Peace Corps—to provide technological assistance for developing countries. A veteran of Geekcorps, Matt Berg, was assigned to run a team of Senegalese and international volunteers who would kick-start the Louma cybercafé.

Initially, Berg was motivated and excited about the prospect of helping new communities get off the ground. He created online accounting software in Wolof that helped local merchants conduct simple transactions. These forms could also be printed out, so that merchants could take them back to their stores, or to their suppliers. Berg also spent a substantial amount of time trying to teach the Senegalese volunteers how to program and how to acquire new IT-related skills.

After seven months, CyberLouma had made a profit of around $2,000. After eighteen months, CyberLouma transitioned from being funded by USAID to becoming a Senegalese-owned and -operated business. When the project shut down for good, however, DFI's Final Report found that CyberLouma had had little long-term measurable impact, and only a small short-term impact: "The main outcome of the evaluation showed that more than 60 merchants opened e-mail accounts, 5 found international commercial partners, 5 illiterate opened e-mail accounts, and 7 traders are using the software KOCC for their daily management."[63]

Berg added that probably his most significant long-term impact was the individual mentoring that he did, and that the entire premise of a business center and new accounting software simply were not sustainable in that type of environment. "The reality of most of these donor-funded cybercafés, really, [is that] maybe at first they have a subsidy for their pricing—but then they become a separate thing that is run by entrepreneurs," Berg said.[64] In other words, a well-funded donor like Hewlett-Packard or USAID can only last so long.

What makes the most amount of money, Berg says, is the same kinds of things that happen at any other cybercafé: kids gathering around, watching YouTube videos and chatting with their friends. Because CyberLouma was the first cybercafé in its neighborhood in the Sandaga area, it became known not only to local merchants, but also to regular people who wanted to get online.

In the end, none of CyberLouma's sister projects got very far, either. There may have been some short-term impact—computerizing a doctor's office, for instance—but there is no evidence that DFI Senegal, which spent hundreds of thousands of dollars per year, had any long-term impact on Senegalese society. At the end of the DFI's second year, John Mack, one of the original architects for the Leland Initiative, was called in to perform an outside audit on the project as a whole. Following his audit, the project was shut down.

Not only was DFI unsuccessful in terms of creating lasting development programs and training on the ground in Senegal, it also had disappointing results in terms of creating any kind of regulatory reform. When DFI was over, there was no better competition through the presence of more fixed-line, Internet, and mobile phone operators than there had been before. Sonatel, still the monopoly telecom, remained and still remains the dominant player.

One former high-level American official in DFI Senegal said that repeated offers of legal and technical training made to the Senegalese government and the ARTP were either politely rebuffed or never acknowledged. This same official said that he was disappointed with how the entire project ended, and admitted that while the DFI "looked good on paper," in the end "we may have been expecting too much."[65]

Both the Joko clubs and the Digital Freedom Initiative failed for the same reasons. The primary reason was that they ultimately were financially unsustainable, and did very little to actually increase connectivity in Senegal. A secondary reason was that while these initiatives were floundering, the Internet became more widespread and cheaper to use all over the country, largely through the explosion of private cybercafés. Finally, both initiatives expected that self-sustaining local capitalism would take over from foreign donors, which did not happen.

WHILE JOKO and the Digital Freedom Initiative were going on, there were also all kinds of other policy-oriented changes happening on the international political level. The international community was beginning to realize that it needed to put into place similar types of measures to advance Internet access in other developing countries, too.

The first such change was the creation of the "New Partnership for Africa's Development," (NEPAD) a strategic political framework document established in July 2001. Five major African heads of state developed NEPAD, including Senegal's President Abdoulaye Wade. Among other development goals, NEPAD established the development of information and communications technologies as one of its eight top priorities.

Later that same year, the United Nations General Assembly passed a resolution on December 21, 2001, that called for the World Summit on the Information Society (WSIS), a massive international conference to be held in Geneva at the end of 2003 as a way to evaluate the state of global Internet access. This conference, in turn, was a specific response to the 2000 United Nations Millennium Declaration, which called for "special measures to address the challenges of poverty eradication and sustainable development in Africa . . . as well as transfers of technology."[66]

In preparation for the 2003 WSIS Geneva conference, there was a regional African conference in Bamako in May 2002, a tech conference of ministers in Dakar in November 2003, and the World Conference of Cities on the Information Society in Lyon, France, on December 4–5, 2003. They all called for the creation of a "digital solidarity fund."

ON EACH OF THESE OCCASIONS, Amadou Top was part of the Senegalese delegation, and took part in informal one-on-one discussions with President Wade. He was even invited to accompany the president en route to the WSIS Geneva conference.

In these private meetings, Top learned what Wade meant by "digital solidarity." The idea didn't gain much public traction until December 11, 2003, when President Wade gave a speech before the WSIS Geneva conference. In this speech, the trained legal and economic scholar appealed to the emotional cores of his fellow delegates and encouraged them to create a fund that would put Africa on an even playing field with the rest of the world:

> What Africa asks for today, is that humanity remembers the great injustices that were done to it and creates the mechanisms that will permit [Africa] to prime the pump that will replace it in the court of the world economy, where it receives less than one percent of investment and where its level of participation in international commerce is less than 1.4 percent. We accept competition, we accept globalization, but on equal footing; we say yes to free trade, but also to honest trade. We must not allow ourselves the freedom of exchange at the same time where, through subversions, we inhibit our products from reaching the markets of developed countries, and are even [unable] to sell them in our own markets. We say "free trade but fair trade."[67]

His speech did not include any instructions to achieve this lofty goal. Nor did he address how it would be different from the previous government and nongovernmental programs that had been attempted in Senegal and other parts of Africa. Essentially, he simply was asking for the countries of the global North to be in "solidarity" with the global South and contribute to a fund so that they could buy more computers:

> In summary, if we pool our money, it would certainly be for us to buy equipment from developed countries that would allow us to connect to the Internet. These countries will eventually get their money back through their own companies and industries, and we will have the equipment, it's what businessmen call a "win-win" situation.[68]

While President Wade rattled off a laundry list of institutions and countries that were behind the Digital Solidarity Fund (DSF), there was a notable absence from the world's most technologically oriented country, the United States.

By the end of his speech, it was up to delegates from various countries to hash out whether or not this proposal could actually become a reality—many Japanese, North American, and European delegates were skeptical of the entire premise. At the conclusion of the conference, WSIS delegates were unable to come to an agreement as to how to actually execute this idea of "digital solidarity." The delegates agreed to study the concept of digital solidarity further and pushed it off until the next WSIS session, to be held two years later in Tunis.[69] However, the municipal governments of Geneva, Lyon, and other cities initially put up over $1 million to fund this idea.

Two days after Wade's speech, Ambassador David A. Gross, the U.S. coordinator for international communications and information policy at the U.S. Department of State, gave a press conference in Washington, D.C., to talk about the American position on what had happened at WSIS. One foreign journalist asked why the United States was not supporting the DSF, and Ambassador Gross took a wait-and-see approach in his answer:

> It was not so much an issue of supporting or not supporting the Digital Solidarity Fund; rather, we thought that these issues, at this time, were premature. We're very comfortable with the way that documents came out which said, in essence, that those countries that would like to go forward and create a Digital Solidarity Fund.
>
> Obviously, countries are sovereign nations, are free to do that which they think are in their best interest, and, of course, we have no interest in interfering with that. The issue was, is there a need for a new multilateral fund to be created, and we thought that the answer to that at this time was premature because, one, we didn't know if existing multilateral

funding mechanisms were sufficient or not. There was no evidence, one way or another. We thought that work needed to be done.

Two is, assuming for the moment something we do not yet know, that is, that a new multilat—that the current funding mechanisms are insufficient, is the better course to tweak the existing funding sources in some way, rather than create a new fund? It's often easier to change existing things, not always, but often it's easier to change than new ones.

Three, why aren't the existing funding mechanisms sufficient, if that's the case? Is it an efficiency problem? Are the wrong people getting the money? Is it just not enough money? What is it? We don't know, and we thought those—that question ought to be answered before we commit to something.

Similarly, who would gather the money? Who would—how—what are the controls about that? Of course, as everyone would agree, when people in the international community gather other people's money, voluntary or not, it's very important that people feel secure about that process. We didn't think we knew the answer to that question at this time.

And then, lastly, we weren't sure how the monies were going to be disbursed because that was a very important question. Are they going to go to people who have already been receiving, but not yet benefiting their people?

That, we thought we'd think, obviously, is a bad idea. I think most people would think that would be a bad idea.

Is it going to be done based on certain criteria to direct it? If so, what are those criteria?

As you can tell, lots of questions, and therefore it was very premature to commit to a new fund. So we are very pleased that the documents say that we all agree that the issue of the need for a fund and the feasibility of a fund ought to be studied, and we'll be very anxious for the result of that study because that will tell us something important, and then it would be ripe to make a decision.[70]

In addition to these profound questions, it's also likely that with the Digital Freedom Initiative just beginning to kick off in Senegal, that the American position, rightly or wrongly, was to try to kick-start indigenous entrepreneurial environments, rather than simply fund technology purchases via cash donations. Still, the DSF moved on without American help.

Continued Changes and Disappointments (2004–2008)

While there was a lot of talk at the WSIS conference, the following month, the Senegalese government finally decided to take substantial action. In January

2004, the government declared that it would finally put an end to the monopoly of Sonatel. This action, of course, came seven years after the company had been privatized. This privatization was a means to make the company more efficient and serve more people—which can only work when a company actually has competitors. Instead, Sonatel became a private monopoly, which is even worse than a public monopoly, as the money flows into the hands of investors and shareholders rather than the state. Worse still, this new private monopoly was and still is owned largely by a foreign and former colonial power, France Telecom. Nevertheless, beginning on July 20, 2004, Sonatel wrote in a press release: "a new era will open up for the telecommunications sector."[71]

However, on July 20, 2004, nothing changed at all. It was as if the government simply picked an arbitrary day as a way to prove that it could implement this policy of demonopolization without any meaningful reform behind it. When the monopoly ended, there was no competitor to step in to try to rival Sonatel. The new era looked a lot like the old one. The fact that neither the government nor the ARTP had the foresight to put this into place beforehand is the most blatant example of incompetence, nepotism, and inexplicable behavior surrounding the deployment of the Internet in Senegal. Further, it took over three years from the formal end of Sonatel's monopoly to the beginning of a formal search for a national operator. No government official has been able to give any kind of coherent explanation as to why this egregious delay took place.

In August 2007, the ARTP organized a tender for a second national operator license (fixed-line, Internet, and mobile). Various telecom companies competed, ranging from Belgian and Saudi Arabian companies to another from Sudan. On September 7, 2007, Senegal announced that Sudan's Sudatel would receive this license, at a price of $200 million, the largest price for a license in all of sub-Saharan Africa, and four times what Millicom had paid to run its Tigo network in Senegal years before. However, one week later the president's IT advisor, Thierno Ousmane Sy, announced that actually it was he and the president's son, Karim Wade, who were charged by the president with making the decision to assign the license.[72] In other words, what should have been a transparent, straightforward governmental process was again marred by nepotism and possibly shady business deals. Some Senegalese IT watchers have even suggested that there is a conflict of interest for Thierno Ousmane Sy in his role as presidential IT advisor and his position on the board of directors of Sonatel.

For his part, Top was furious about the way the Sudatel deal took place and spoke out against it in his *Batik* newsletter:

> In this atmosphere the complaints by APIX [a government entity to promote foreign investment in Senegal] against Senegal's 162nd place out of 178 countries in the World Bank "Doing Business 2008" report, are ridiculous and pathetic. In effect, what's the point of putting in place a

framework that conforms to the highest international standards when, in fact, the daily business reality is based on something else entirely? What has just happened sends a bad signal to foreign investors with major projects in Senegal and who wish to be able to operate in our country without this sort of pressure. It is therefore high time for those at the core of the state apparatus, who wish to bring us out of its protective lap, to behave like real capitalist entrepreneurs who take risk rather than using and abusing their positions of power. Doing business, yes, but in the right way.[73]

Sudatel, for its part, had originally promised to begin its "Expresso" mobile phone service in Senegal in January 2008. Expresso was then delayed to May 2008, then put off yet again, with no new start date. Service finally began on January 13, 2009.

Regardless of whether or not these delays were intentional, it's clear that the main beneficiary has been Sonatel, which remains one of the most profitable companies (likely the most profitable) in all of Senegal. In 2007, the company had a net profit of more than $310 million—about 2 percent of the entire Senegalese economy. With no real competitor in landline or international calls, and as the dominant player in mobile phones, it's likely that these profits will only continue to rise.[74]

BUT THE DEMONOPOLIZATION and liberalization of the Senegalese telecommunications market is not the only part of the story that's been marred by dysfunction and incompetence. President Wade's plan to create a digital solidarity fund also has not worked thus far.

Discussions of the Digital Solidarity Fund that began during the 2003 WSIS Geneva conference were continued two years later at the 2005 WSIS meeting in Tunisia. Again, President Wade gave a lengthy and impassioned speech about the necessity of digital solidarity, without explaining how to make the concept work. However, by the end of the conference, delegates from various countries had agreed to make the Digital Solidarity Fund a voluntary fund, instead of mandatory, as had been suggested two years before, and came up with a mechanism by which the fund could accumulate contributions. Countries, regions, and cities could contribute in two ways: through direct donations and through the "one percent principle."

The idea behind this principle was that governmental entities or corporations could impose a 1 percent clause as part of the cost of doing IT-related business with them. For example, if the City of Geneva had a pending IT contract with Microsoft for $5 million, then 1 percent of that amount ($50,000) would go to the Digital Solidarity Fund, and would be deducted from the profit margin of Microsoft. The theory was that if enough government entities imposed this requirement, then substantial sums of money could be easily collected. The organizers like to call this a "voluntary contribution," when in fact it essentially is not voluntary and effectively is, for better or worse, a tax on the corporations.

Within two months of the conference's conclusion in 2005, the Digital Solidarity Fund was formally established in Geneva, Switzerland. (Amadou Top was made a vice president of the DSF until late 2007.) Originally, as Lyon had been one of the first cities where discussions of the proto-DSF had been held, Lyon would be the executor of the funds and would plan their use, while Geneva would be the seat of the funds. However, due to a bureaucratic dispute, Geneva went ahead and created not only a financial institution to house the funds, but also an executive body to distribute that money, headed by Alain Clerc, the co-president of the "Fondation du Devenir," a Geneva-based NGO that aims to improve "quality of life" around the world. In 2006, Lyon set up its own sister organization, the World Digital Solidarity Agency (WDSA), headed by Jean Pouly.

During the first four years of its existence, the DSF received approximately €8 million in one-time donations from cities and countries around the world that wanted to show their support for the cause—and pledges for approximately €2 million more. However, the donations were hardly proportional based on each country's ability to pay. For example, France donated €300,000 (around $384,000), which was just about as much as was donated each by the Geneva City Council and the Dakar City Council, and slightly higher than the amount donated by the entire Republic of Ghana.[75] These single pledged donations came from cities (Geneva, Dakar, Lyon, Malaga, and Santo Domingo), regions (Rhône-Alpes), countries (Algeria, Nigeria, Mali, Morocco, France, Mauritania, Kenya, Guinea, Equatorial Guinea, and Senegal), organizations (International Organization of Francophonie), and companies (StratXX).

After four years of existence, the DSF is looking more and more like a simple charitable organization that relies on private donations, and again, not one that has a consistent, innovative, and long-term source of funding. Indeed, to date, only five cities, a few corporations, universities, and the Republic of Senegal and have adopted the "one percent principle" and, as of late 2008, have only collected a total of €90,000 ($115,000). It is also possible for the Digital Solidarity Fund to receive donations from individuals who wish to support this noble cause. As of November 2008, the DSF has received three such donations, totaling €120 ($155).[76]

However well-intentioned the DSF may be, the organization is woefully wasteful, or at least grossly mismanaged. By its own account, in 2008 the DSF budgeted over $430,000 worth of project investment, which required nearly triple that amount in operational costs for office space in a Geneva villa, five full-time staff, assorted logistics, and travel costs. Jean Pouly, the director of the Lyon-based WDSA, the DSF's sister organization based in Lyon, said that this is largely the result of frequent and unnecessary trips made by Alain Clerc to various international locations for conferences and to get general promises of support for the DSF. As a result of this disagreement, at the November 2008 digital solidarity meeting in Lyon, the director of the DSF, Alain Clerc, was asked to step down by members of the WDSA—a move that he initially refused.[77]

Because of this dispute, an emergency meeting of the DSF was held in Bamako, Mali, in January 2009, the first time such a meeting has taken place on such short notice, and the first time one has taken place in the developing world. Ababacar Diop, President Wade's special advisor on digital solidarity, considered this meeting to be a complete "re-orientation" of the entire premise of digital solidarity.[78]

The Digital Solidarity Fund has been used to fund an assortment of tech-related projects in different countries around the world, including Colombia, Congo, Tunisia, Burundi, Burkina Faso, and others. However, none of these projects have had any demonstrable long-term or sustainable effect. Some of the programs—such as the e-waste facilities in Africa—are supported with additional funds donated directly from corporations (in this case, 87 percent of the e-waste programs are funded by HP). Furthermore, the pilot programs in Burundi, Niger, and Burkina Faso were halted in 2008. In short, despite the fact that the DSF's literature speaks of "innovative financing"—also known as the "one percent principle"—only a very minute portion of the DSF's money comes via those means. In short, the DSF has now formally existed for over six years and has very little to show for its efforts.

Conclusions (2009–Present)

In the end, while countless organizations and government programs have seen the great potential for the Internet in a country like Senegal, nearly every single hope has ended in disappointment. For the most part, Senegal's level of technological advancement—while relatively superior to its sub-Saharan African neighbors—is still far off from where it could be. As a result, its potential has largely been squandered.

This problem began long before the Internet came to Senegal in the 1990s. In 1982, the French government tried to promote its "Minitel" project, a sort of proto-Internet that ran on proprietary computer terminals. It partnered with Apple (which provided its Apple II computers) and MIT's Seymour Papert and Nicholas Negroponte to help Senegalese students learn how to use the LOGO programming language. While this project may have proved Papert's theory that kids anywhere in the world can gravitate toward and understand computer technology at a young age very easily, the project ultimately failed. Within a year, Papert and Negroponte had both quit, and the French government pulled the plug soon after. As a 1983 piece in MIT's *Technology Review* presciently concluded: "Besides, altruism has a credibility problem in an industry that thrives on intense commercial competition." Negroponte and Papert are now in charge of a similar, albeit more ambitious, program, the "One Laptop per Child" project, and OLPC is targeting Senegal yet again.[79] This concept of providing computer and technological help via charitable means is unlikely to make any significant inroads a quarter-century later.

The reason why more Senegalese people aren't online, much less using computers, has little to do with their intellectual capacity. Rather, it reflects much more basic problems. First and foremost, the majority of Senegalese people (60 percent) cannot read or write in French, nor in any of Senegal's indigenous languages. Worse still, despite the fact that in 2005 Microsoft committed to creating a version of Office in Wolof, Senegal's largest language, Wolof literature and literacy is essentially nonexistent.[80] Far fewer people can read and write in Wolof than French. This concept may seem simple and obvious, but it's often overlooked: you can't use a computer (or the Internet) if you can't read. In Senegal, that means 8.4 million people, or about 60 percent of the population, is offline automatically.

By comparison, in both Korea and Estonia, nearly everyone can read. Even in Iran, literacy approaches 80 percent. Estonia achieved very high levels of literacy over a century ago, and has been able to reap the rewards of an advanced political and economic system, and as such, it was ready to take advantage of Internet technology once it was presented to them. Senegal needs to overcome this extremely basic but fundamental truth before it can hope to get the majority of its citizens online.

Beyond literacy loom the problems of unemployment and low economic activity. While it is true that many people in the capital city, Dakar, have both physical and financial access to a cybercafé, the truth is that throughout the country nearly half of the workforce is unemployed. In a country where even the most basic used laptops cost in the hundreds of dollars—10 to 20 percent of the average annual salary (equivalent to $4,500 to $9,000 annually for the average American consumer), it's no wonder that most people cannot afford a computer. Even if OLPC does manage to get its retail price down to $100 per machine, that would still represent a substantial portion of the average African's income.

This economic imbalance explains why so many Senegalese access the Internet from public cybercafés—a shared model that has been shown to work much better than the global North model of one computer per person. It keeps the costs of access down. While lots of Senegalese people, especially in the Dakar region, have near ubiquitous access to cybercafés at affordable prices (no more than $1 per hour), the fact remains that a significant portion of the country does not have access to the Internet because telecenters and cybercafés remain concentrated in urban areas. Even those who are capable and could afford to pay it, in many cases, are simply out of reach of a cybercafé or a DSL line. This situation would surely be remedied if Sonatel did not almost completely control the residential Internet market—there are no major cable or DSL rivals to speak of—so Senegalese citizens are at the mercy of where Sonatel decides to operate. All of the small handful of other companies have to buy access from Sonatel, which, via its parent company, France Telecom, owns part of the SAT-3 undersea data cable, the country's main way to access international networks.

A Senegalese ISP buying access from Sonatel has no choice but to fork over $1,316 per megabit per month. While this price is one of the lowest on the entire SAT-3 network, that ISP is legislatively forbidden from buying any supplemental access via satellite, which, again, further keeps the price of Internet access for ordinary consumers artificially high. Relying largely on one channel for international access means that prices for Internet access will continue to be artificially high in Senegal and in much of Africa for the foreseeable future. It costs even more for landlocked African countries that do not have their own undersea-cable landing station, like Senegal's neighbor to the east, Mali, which has to buy access from Sonatel and transmit it overland via fiber optic cable. Again, monopoly conditions oddly dictate that the current market rates for overland data access from Dakar, Senegal, to Bamako, Mali, cost twice as much as the cost of getting that data from Europe to Dakar.[81]

While it is true that Sonatel has dropped its rates for both telephone calls and for Internet access in the country over the last few years, even the most basic DSL plan still costs approximately $40, which is more than what it costs in most European, Asian, and North American countries. Still, while Senegal is lucky to have the level of international bandwidth that it has (nearly three gigabits), Estonia had about half that amount in 2005, while having only one-tenth the population of Senegal.[82]

On the other hand, Senegal is blessed with a president who seems to pay a lot of attention to the Internet and considers it a significant part of his national and international policies. Not only does he talk a lot about these issues, but also his influence is global. After all, he successfully parlayed his enthusiasm into a new international organization, the Digital Solidarity Fund. Despite Wade's best intentions and lofty language, the DSF has been a colossal waste of money. European officials like Jean Pouly, the director of the World Digital Solidarity Agency, are proud to talk about how this is a truly African initiative, and thus is different than its predecessors.[83] While it is true that the concept of the DSF originated from an African thinker and politician, it speaks volumes that both the DSF and the WDSA are based in Europe, not on the African continent. There is no reason to believe that these "innovative" financial mechanisms will be any more successful than other similar types of charitable instruments that have come and gone before them.

Already, though, there is a means for Senegalese people and other Africans to begin to get a taste of the Internet beyond what they may have already seen on computers in cybercafés—via cell phones. Africa is the fastest growing mobile phone market in the world. The continent experienced an annual average mobile usage increase of 58 percent between 1999 and 2004. This happened because the Senegalese government (and others across Africa) simply got out of the way and allowed private companies to compete with one another. As a result, mobile penetration in Senegal approached 50 percent in 2008, and has since surpassed that level.[84] While using a mobile phone is much simpler and requires significantly less knowledge than a computer and the Internet, it stands to reason that Internet access via mobile phones will grow significantly in places like Senegal. Further, the Senegalese and their Africa counterparts have done nothing but provide the legal means for companies to come operate in their country. Internet access will only come once a similar fluidity in the market can take place.

The Senegalese government should have observed that, just as telecenters exploded in the 1990s, if it wants to encourage Internet access in the same way, then it should simply study its own telecom history. In other words: privatize the industry in a real, meaningful way, and let economic forces drive the prices down.

Meanwhile, Amadou Top continues to run OSIRIS as before, and he continues to write his issues of *Batik*:

> At the end, despite the official existence of three operators (Sonatel, Sentel, and Sudatel) on the Senegalese telecommunications market—termination segments and traffic collection on fixed lines, national and international traffic, data transmission, capacity location, the IP market, access to special services, and necessary signaling for international roaming—still make Sonatel a monopoly that largely dominates mobile telephony, of which it controls 79.5 percent in terms of traffic. In conclusion, the fruits of liberalization of the telecommunications market haven't held the promise of flowers. Freedom of choice, betterment of quality and diversification of services and a lowering of fees, especially national, are still hypothetical for Senegalese whose only certainty is having lost their national operator into the hands of great international capital, that's how liberalization doesn't rhyme with liberation![85]

TIMELINE

1947: Amadou Top born in Senegal

1996: Senegal connected to the Internet

February 1997: Senegalese government website launched

July 1997: Sonatel privatized

1998: National election registries online

March 1998: OSIRIS created

January 2001: Joko created

August 11, 2001: Multimedia Caravan launched

December 2001: ART created

March 4, 2003: DFI launched at White House

December 10, 2003: Geneva conference; DSF proposed; UNESCO donates 400 million CFA francs (around $800,000) to create cybercafés and other community centers across Senegal

July 19, 2004: Sonatel demonopolized

March 14, 2005: Digital Solidarity Fund created

3

Estonia

After having dug to a depth of 100 metres last year, Scottish scientists found traces of copper wire dating back 1,000 years and came to the conclusion that their ancestors already had a telephone network more than 1,000 years ago.

Not to be outdone by the Scots, in the weeks that followed, English scientists dug to a depth of 200 metres and shortly after headlines in the newspapers read, English archaeologists have found traces of 2,000-year-old fibre-optic cable and have concluded that their ancestors already had an advanced high-tech digital communications network a thousand years earlier than the Scots.

One week later, Estonian newspapers reported the following:

After digging as deep as 500 metres in Narva, Estonian scientists have found absolutely nothing. They, therefore, have concluded that 5,000 years ago, Estonia's inhabitants were already using wireless technology.

—Early twenty-first-century Estonian joke, author unknown

Tallinn

February 8, 2007

It's impossible to walk through Tallinn, the medieval-era capital of Estonia, without noticing centuries-old dark gray stone walls and protective arches surrounding the old city. Cobblestone streets line the interior city roads and connect the ancient fortresses, twelfth-century church, and converted eighteenth-century palaces along the hilltop anchoring the city center. However, in the city parks, just as in many other parts of the country, it's equally impossible not to notice the large orange and black signs. These industrial-grade metal road signs proclaim: "wifi.ee / Tradiitaa Interneti leviala / Area of Wireless Internet." The signs are perched at every gas station, and in many public parks and other public spaces. Smaller versions of the sign, in sticker form—not unlike credit card stickers in the windows of restaurants—adorn the glass doors and windows in nearly all pubs, hotels, and cafés.

One of these cafés is the two-story Kohvik Moskva, or Café Moscow, a modern, trendy spot that overlooks Vabaduse Väljak, or Freedom Square. This square marks one of the boundaries of Vanalinn, the old city, and is a short walk from the spot where Soviet bombers destroyed the residences of some twenty thousand Estonians in 1944.

Kohvik Moskva is replete with white and black leather chairs, smartly dressed wait staff, soft electronic music, and expensive cocktails. It easily could otherwise be in any North American or European city. Seated at one of the tables is Veljo Haamer, at forty years old still wide-eyed and baby-faced. His blue eyes always seem to be looking at something farther down the horizon, and the upper right side of his temple is marked with a three-inch-long scar from a bar fight many years ago. He's the man behind those metal signs. Essentially single-handedly, he has preached the gospel of WiFi, café to café and pub to pub, all across the country. Because of him, the vast majority of Estonian hotspots are free—and even when they're not free, the price tops out at €1.5, or the equivalent of $2 US, for twenty-four hours. This ubiquity of free connectivity is in stark contrast to nearly everywhere else in Europe, where one hour of WiFi at the Helsinki airport, just a two-hour ferry ride to the north, costs about €8. Haamer has also installed WiFi on a regular bus route to Riga in neighboring Latvia, on the ninety-minute ferry to Hiiumaa, one of the islands off of Estonia's western coast, and on Tallinn's suburban electric train line. All are free.

With such a high level of connectivity, the entire country has effectively become Haamer's office. On this day, he's working comfortably in one of Kohvik Moskva's large black leather chairs, and looks up from his laptop.

"Just three years ago, I was the only one here with a laptop," he muses with a grin.[1]

A quick glance shows that perhaps half a dozen patrons out of the fifteen seated—mostly university-age kids—are clacking away the afternoon.

TODAY, Estonia is a country that has fully embraced the Internet in a way unparalleled anywhere else on the globe. The World Economic Forum ranked it as number one in the world in 2006 in the categories of "Laws relating to ICT," "Availability of online services," and "ICT use and government efficiency." In Estonia, 97 percent of banking transactions are done online, and the majority of people pay their taxes online.[2] It became the first country in the world to allow its citizens to vote via the Internet in 2007, and will allow its citizens to vote via mobile phone in 2011.

Estonia gives each citizen and legal resident a national identity card (roughly the size of an American driver's license), which has an embedded microchip inside. Each person has access to a "citizen's portal," where any

Estonian resident can access university admissions exam results, maternity payments, and other interactions with the federal government. Even the president, Toomas Hendrik Ilves—a technology enthusiast in his own right—has taken notice, and frequently underscores this point in interviews. He was quoted saying as much in the English-language paper the *Baltic Times* in 2007:

> In Estonia you can use your computer anywhere and you have free WiFi. In the worst case, you pay one Euro for 24 hours. In most of Europe you can pay up to 8 euros for 30 minutes.
>
> When I moved in Tallinn from one apartment to another and asked for an Internet-service, the company told me that they could come on the very same day between 2–3 pm and how did this time slot suit me? When I moved to Brussels, I had to wait seven weeks from when I applied for the Internet service![3]

When Estonia became independent from the Soviet Union in 1991, the West largely ignored it. It had a barely functioning telephone system, a nonexistent banking system, and a government that had been in exile for decades. In 1987, the nation's gross domestic product per capita was optimistically estimated at $2,000. That was just higher than Senegal's GDP per capita today. At that time, it was also about one-seventh the value of neighboring Finland, according to historian and former Estonian prime minister Mart Laar.[4]

As Laar wrote in a 2007 article published by the Heritage Foundation:

> The end of communism had created real chaos in the country. Shops were completely empty, and the Russian ruble no longer had any value. Industrial production declined in 1992 by more than 30 percent—more than during the Great Depression of the 1930s. Real wages fell by 45 percent, while overall price inflation was running at more than 1,000 percent and fuel prices had risen by more than 10,000 percent.
>
> People stood in lines for hours to buy food. Bread and milk products were rationed. Because there was no gas for heat, the government planned to evacuate much of the capital of Tallinn to the countryside. The only "institution" in Estonia that seemed to work was the informal market.
>
> Estonia was absolutely dependent on Russia, which accounted for 92 percent of Estonian international trade. Estonia had little that it could sell on world markets. The Soviet command economy had ruined Estonia's environment, and the infrastructure was in catastrophic shape. For most foreign experts, Estonia was just another "former Soviet republic" with not much hope for a better future.[5]

As Linnar Viik, a longtime IT advisor to the Estonian government, who is often called the "Father of the Estonian Internet," also remembered: "When the first democratically elected president of Estonia moved into his office, there were six telephones on his desk—three red and three green. All of these could take incoming calls, but not a single one could be used for dialing out."[6]

Yet somehow, in less than twenty years, the country created a stable currency, the Estonian kroon, and by 2004 joined the European Union and NATO. (It was the first post-Soviet republic to create an independent currency and abandon the Russian ruble.)[7] Real wages, income, wealth, and property values all continued to steadily rise. While the Internet was beginning to boom in Silicon Valley in the late 1990s, Estonia began a national project to wire every school across its flat, forested territory. At the time, most Estonians had barely even heard of the Internet. Yet, bringing the Internet to Estonia—a technology that was not well understood outside a few visionaries—was going to cost millions of tax kroons.

Within a decade, the government had created an online system of debating pending bills in the cabinet, and citizens could pay their parking meters via cell phone. It was this same country that spawned the Internet phone company Skype, which, according to a 2010 telecommunications industry report, is the "largest provider of cross-border communications in the world."[8] All of these technological innovations, as well as countless others, have emerged from this nation of 1.3 million people. In short, Estonia has become a proud outpost of economic and technological prosperity on the northeastern frontier of Europe.

New York City

July 16, 2002

Just a short walk from Times Square, Bryant Park is a place where many New Yorkers go to escape the hullabaloo of midtown Manhattan. With its dark green tables and matching chairs, it echoes Parisian parks. Earlier that month, NYCwireless.net, a nonprofit organization, had just opened up one of the first public wireless Internet access points in the United States. The *New York Times* reported on Bryant Park's WiFi debut and quoted one of its first users, a Brooklyn photographer named Oren Eckhaus. He was astonished at seeing the confluence of the Internet and a natural setting:

> "I'm surrounded by all this technology, and this leaf falls [on my keyboard]—that is so amazing," Mr. Eckhaus said, sitting in the shade on a bench in Bryant Park last week. "Nothing like that can happen at home, except the coffee can spill on your computer."[9]

Across the north Atlantic, Veljo Haamer read about this development from his apartment in Tallinn. Earlier that year, he had first been exposed to WiFi while working in a local computer shop. When Haamer first heard about NYCwireless's work through technology websites, he immediately planned a trip to the United States to see the network for himself. He wrote to Terry Schmidt, one of the founders of NYCwireless:

> hi terry, we are small wifi activist in small european country called estonia. we have 1.5 mil people and 30 wifi covered areas. we come to see nyc wireless activist.[10]

The e-mail went on, in broken English, to explain that he and a friend would be arriving in New York within a few days and would very much like to meet with someone from the group. Upon receiving the e-mail, Schmidt sent another member of NYCwireless, Jacob Farkas, who worked near the park. He agreed to meet Haamer, mainly out of a sense of curiosity. Because Jacob Farkas's own family was from Ukraine, he thought there might be a certain sense of post-Soviet fraternity between them. But more so, he was astonished that there was any WiFi at all in Estonia. At the time, the only other organized wireless projects that were going on in the United States were in the tech hubs of San Francisco, Portland, and Seattle.

"I didn't know of any in Europe, let alone Estonia," Farkas recalled. "I didn't picture Estonia to be that forward. I was envisioning rural Ukraine, rural New York at best. I didn't envision that there was a dedicated group of people [bringing wireless to Estonia.]."[11]

As soon as they met and Haamer confirmed that the WiFi indeed worked as advertised, he immediately opened up MSN Messenger, an online chat program, and began sending messages to his son Kris and ex-wife Liivi back in Estonia.[12] For Haamer, who had set up a couple dozen WiFi hotspots around Estonia by that point, the concept of a public outdoor space offering "free communication" had never come to mind.

Haamer peppered Farkas with questions about the specifics of the park. How many antennas did the organization need to cover its nearly ten acres? How did the team manage to hide the antennas so that they stayed out of sight? Farkas did his best to answer, saying that it was important for the park to not only have a network, but to have a network that provided full coverage for the entire area. Further, the park authorities didn't want to mess with the aesthetic look of the environment, so they hid the antennas off of the newspaper kiosks and in trees, and did their best to blend them in with the existing setting.

As soon as he could see what was possible, Haamer felt newly invigorated. Before coming to New York, Haamer had been running his website, WiFi.ee, full-time since March 2002, and brought free WiFi to Tallinn Airport in April of the same year. One month later, he had created the first WiFi.ee signs and stickers as a way of showing the presence of WiFi. It was a much simpler way of finding wireless Internet access than walking around with an open laptop to find such access points (or as some American geeks had dubbed it: "warwalking"), and the subsequent, short-lived tech phenomenon of marking such spots with sidewalk chalk ("warchalking").

A Pre-WiFi Veljo Haamer (1967–1989)

Viljandi County, Estonia

August 23, 1989

The sun shone down on Veljo Haamer as he stood on a road in southwestern Estonia surrounded by forest near the Latvian border. At just less than six feet tall, Haamer held hands with his wife while his short blond hair rippled in the late afternoon summer breeze. When he turned his eyes north, toward Tallinn, he could see people all the way to the horizon. When he looked south, toward Latvia, he could see people all the way to the horizon. Everyone around him was singing folk songs, and also songs of liberation—including one called "Sunrise." This Baltic chain was a million-person link that stretched hand-to-hand from Tallinn in the north to Vilnius, the capital of Lithuania, in the south.[13] The Haamer family had been bused in, assigned to this part of Estonia for one day to form their part of the chain. They sang in a daring act of defiance against the Soviet regime, demanding their freedom from the Soviet Union. The Berlin Wall fell nearly three months later.

But Haamer had other reasons for celebration. At the time, he was working for the computer sales company Korel, which he had helped found one year earlier. Business was starting to grow and he and his partner were on a path to an average annual profit of 100,000 Estonian kroons, or nearly $12,000.[14] This approximated five times the average salary in Estonia at that time, and Haamer was only twenty-two years old. He had a wife and young son, and provided for them, living and working in Tartu as a computer retailer. What a way he'd come from being a young boy from a rural mining town in the northeastern part of the country. Now he was watching the future become even more real—as political, economic, and social freedom proliferated. Life was good, and it was about to get better.

Haamer was born in 1967, during the peak of the Soviet occupation. He grew up in a stodgy apartment block in Kohtla-Järve, a small, quiet mining town outside Narva, in the northeastern part of the country close to the Russia border. His father was a local bus driver, and his mother took care of the house and the family. Haamer attended high school from 1983 through 1986 at Nõo Reaalgümnaasium, or the Nõo Secondary School, roughly 140 kilometers (about 80 miles) from Tartu, Estonia's university capital and second largest city.

Nõo Secondary School was then (and continues to be) a specialized and competitive school for students on an advanced math and science track—one of the best in the country. Many students, including Haamer, came from other parts of the country to go to school there. They were to be the new generation of the technologically minded military elite. These students, brought from around the country, would stay at the school's boarding facilities. Haamer, however, rented a flat at the age of sixteen, to live with his girlfriend, Liivi Kruus. The pair married after graduating high school.

Nõo Secondary School was among the first high schools in Estonia to have a computer. Initially, the school had large IBM mainframes that operated on punch cards with a few terminals in the classroom. By the end of Haamer's high school tenure, the school had obtained a few IBM XT computers, the first IBM personal computer (then also known as a "microcomputer") to have a built-in hard drive.

During physics class, Haamer and his classmates had long and dry programming lessons. Mostly the assignments consisted of learning to make simple mathematical computer programs that could calculate the Pythagorean theorem. Haamer, uninterested, did the assignments half-heartedly. He and his best friend, Ivo Kivinurk, passed the time by playing the Estonian version of tic-tac-toe, with a five-by-five grid.[15]

Decades later, many of Haamer's classmates would become the technological leaders of the country. One of them, Martti Raidal, became one of the youngest doctoral students to graduate from a Finnish university, and now is a particle physicist based in Tallinn. Another, Heikki Kübbar, headed the Internet banking program at Hansabank (now called Swedbank), one of the largest Estonian banks. A third, Mait Rahi, now runs Arvutid (formerly MicroLink), the largest computer manufacturing company in the country. A fourth classmate, Tarvi Martens, went on to head the nation's digital ID card project and the e-voting system for the National Electoral Committee.[16]

But even though programming didn't instantly captivate Haamer, once he could explore computers on his own time, his interest grew. Haamer and a handful of his friends used to meet up at the local train station once a week at 7 P.M. to make the short ride over to the Tõravere Observatory, six kilometers from their school. This observatory, built by the Soviets in the early 1970s, was constructed on a forested hill outside the city to avoid light pollution.

The father of Haamer's friend Martin Eermäe worked at the observatory. This connection enabled Eermäe to bring his friends, including Haamer, to spend some time not only with the observatory's telescope, but also on the personal computer in his father's office. Computers at this time were rare, and personal computers were even rarer. In fact, having a personal anything under the Soviet Union was nearly impossible. The half dozen or so boys would each take turns at the telescope, turning their attention to the only thing beyond their borders that wasn't restricted. Meanwhile, another boy sat in front of the computer and explored new things like databases and spreadsheets, and read astronomical texts directly on the screen. Even though the texts used foreign, complicated scientific words, the simple act of using a small "mouse" to scroll text up and down the screen was endlessly fascinating.

The office, which was also well heated, served as an extracurricular hearth for the group. Staying near the telescope was cold, as the evening air rushed in through the open rooftop that allowed the telescope lens to point skyward.

Some huddled over the computer screen, while others lounged on chairs and chatted. At various points during the evening, they would rotate and each would get a turn at the telescope and the computer. But Haamer was the one who bathed most frequently in the glow of the green and black screen.[17]

Estonia: From Invasion to Innovation (1219–1991)

Estonia is a very small country—in American terms, it would be the size of Vermont and New Hampshire combined. It measures roughly 200 miles top to bottom and approximately 250 miles east to west. The name of its capital, Tallinn, is believed to come from the combination of two Estonian words: *Tal,* meaning "Danish," and *Linn,* meaning "city" or "castle." Danish holy warriors invaded the northern and western parts of the country, settling the capital in 1219.[18] Meanwhile, the German army invaded and colonized Livonia, present-day southern Estonia and Latvia. By the sixteenth century, Protestant Lutheranism had begun to seep from its roots in Germany all the way toward and into Livonia.

Various other foreign powers were carving up parts of Estonia by that time as well. The Russian and Polish-Lithuanian armies both jockeyed for power, but the Swedes proved the dominant force by the early seventeenth century. The Swedes imposed Lutheranism, their own official religion, onto the Estonian population. Martin Luther advocated that people should be taught to read scripture in their own language, and as a result of Lutheranism's spread, literacy in the Estonian language spread throughout the land alongside the religion. Also during this time, the Swedes founded the second university within their dominion. Tartu University, then known as Academia Gustaviana, was founded in 1632.

But by the early eighteenth century, an anti-Swedish alliance began to form, consisting of Danish, Russian, and Polish troops. When Polish troops began attacking from the south, Russian forces invaded from the east and proceeded to destroy many city structures across the Estonian territory. In 1710, the Great Northern War was over and Estonia lay firmly in Russian hands.[19] Estonia, however, had been decimated during the war and an ensuing plague. The population had been reduced from approximately 400,000 people to nearly 150,000. (By comparison, the present-day population of Estonia is nearly 1.3 million people.)

From 1800 to 2000, Estonia underwent rapid change in every way imaginable: there were agricultural reforms and high levels of improvement in literacy and education, as well as rapid industrialization and significant technological change. In the latter half of the nineteenth century, communications and transportation technology grew rapidly—the telegraph reached all major Estonian cities in the 1860s, with Tartu and Tallinn receiving telephones by the 1880s, and in 1896 the national railway was completed.[20]

In the early 1900s, particularly in the aftermath of the 1905 anti-tsarist revolution, Estonia developed a young literary and artistic movement that began to explore inherently Estonian themes. As a leading Estonian historian, Toivo Raun, wrote in *Estonia and the Estonians:*

> In 1906, [this new movement's] leading figures—Gustav Suits, Friedebert Tuglas (1886–1971), and Johannes Aavik (1880–1973)—were all 25 years of age or younger. Infected by the all-empire liberation movement, this new generation sought to emancipate Estonian culture from its narrow Baltic world. Suits argued that only through the assimilation of the best of European culture could the Estonians create a modern culture of their own.[21]

But Russian forces continued to occupy Estonia until 1918. By that time, the Russian Revolution had completely enveloped the country. Though Bolsheviks tried to institute communism across the land, Estonia declared its independence from the newly born Soviet Union on February 24, 1918. Within three months, Great Britain, France, and Italy recognized Estonia's independence, and in May 1918 the Soviet Union formally gave up its sovereignty over Estonia.

This independence proved relatively short-lived. On August 23, 1939, Stalin and Hitler signed the Molotov-Ribbentrop Pact, which partitioned Eastern Europe into zones of influence—the Baltic states fell into the Soviet sphere rather quickly. (This was why Haamer and so many other Estonians conducted their Baltic chain on August 23, 1989: the fiftieth anniversary of the pact.) The Red Navy blockaded Estonia, Soviet bombers flew over Estonian airspace, and the Red Army massed along the border. Just over a month later, a treaty was signed in Moscow that forced the Estonian acceptance of Soviet military bases in the country. The Red Army and Navy outnumbered the Estonian Army and Navy several times over. By the summer of 1940, Estonia was entirely subsumed into the now more mature Soviet Union. In the first year of occupation, 60,000 Estonians were deported to Siberia—10,000 in one night alone in June 1941.[22]

As the Estonian State Commission on Examination of the Policies of Repression wrote in *The White Book: Losses Inflicted on the Estonian Nations by Occupation Regimes (1940–1991):*

> From 17 June 1940 the Republic of Estonia was completely occupied land and had factually lost all characteristics of an independent state. All the political, economic, and other rearrangements that followed were carried out under the dictate of the Soviet Union, pursuant to the orders of the Embassy of the USSR, the Soviet military leadership or Andrei Zhdanov, the special commissioner of the Soviet Government, who arrived in Tallinn on 19 June 1940. On the same day, Zhdanov met the President of the Republic and informed the President of his plans.[23]

During World War II, Nazi Germany took over Estonia from 1941 to 1944. Then, after the defeat of the Nazis, the Soviets reclaimed Estonia and occupied it ruthlessly. Eighty thousand more Estonians were sent to Siberia and Central Asia between 1945 and 1953, and hordes of Russians and Soviet officers moved in as part of an increased Russification program. Prior to World War II, Estonia was 90 percent ethnic Estonian, but after the war that figure was down to 60 percent.[24] Despite the horror that was inflicted upon Estonia as a nation, by the end of the Soviet occupation the Estonian Soviet Socialist Republic was, relatively speaking, the most well-off and best educated republic in the entire USSR.

The Kremlin made studying the liberal arts or the social sciences nearly impossible by limiting the number of courses that could be offered in those areas. Instead, the Soviet regime wanted Estonians, who had historically been better educated than their other Soviet-occupied counterparts, to study engineering and computer science. Indeed, Haamer himself was tracked toward physics and computer science.

Given that Estonia has very few natural resources—aside from oil shale and timber—it made sense to focus the country's efforts on science. The Kremlin set up the creation of the Institute for Cybernetics in Tallinn in 1960 (the first of its kind in the entire Soviet Union), and the Institute of Astrophysics and Atmospheric Physics at Tõravere (outside Tartu) in the 1970s.[25] Each institute had a military component, in addition to their stated academic goals.

While Moscow wanted the Estonians to become the Soviet Union's scientific elite, the Politburo also wanted to control intellectual activity. During the years of Soviet occupation, foreign products—much less, information from non-Soviet sources—was extremely difficult (and often dangerous) to come by. It was virtually impossible for Estonians to get an education outside of the country; often they had to get special passes to travel to the border regions, including the regions along the northern and western coasts. However, information did begin to trickle into Estonia during the last years of the Cold War through Finnish-language television and radio stations that were broadcasting just across the Gulf of Finland. Finnish is linguistically similar to the dialect of Estonian spoken in the northern regions, so understanding the broadcasts was not difficult.

Despite Moscow's efforts, Estonia had substantial telephone connectivity, with an average of twenty lines per one hundred people, double the rate of the rest of the USSR. While *perestroika* took hold across the Soviet Union to reform the stiff socialist policies—the first commercial bank in the entire USSR was opened in Tartu in 1989—Estonia's demands for freedom became louder and more pronounced. In addition to the Baltic chain protest, 1989 also saw the declaration by the Estonian Supreme Soviet establishing Estonian as the official language. Finally, in an act of great defiance, 1989 was the first year in many decades that the Estonian blue, white, and black tricolor flag was raised. Protestors chose the eleventh-century tower atop Hermann Hill (the highest point in the capital city, Tallinn) to celebrate Estonian Independence Day on February 24, 1989.[26] In November, the Estonian Supreme Soviet declared that Estonia's incorporation into the Soviet Union was illegal and invalid. On March 3, 1990, an overwhelming majority of Estonian citizens voted for independence.

Despite the Soviet Union's best efforts to industrialize the country and provide consumer goods, the Estonian economy was in shambles toward the end of the Cold War. According to the Estonian State Commission, the per capita gross national product of Estonia was 4,030 rubles in 1989—an amount equal to $640 US, making Estonia poorer than 110 other countries in the world.[27]

Yet, studies at the time suggested that Estonia was far better off than its neighbors. As the Estonian scholars Tönu Parming and Elmar Jarvesoo wrote in *A Case Study of a Soviet Republic: The Estonian SSR* (1978): "The Estonian SSR is the most industrially advanced and urbanized of the Soviet republics and also, perhaps logically, has the most Western life style and the highest standard of living."[28]

The World Bank, which accepted Estonia into its body in 1992, agreed, writing in its 1993 study of the country:

> As a result of its economy's higher efficiency, Estonia enjoyed the highest standard of living among the republics of the former Soviet Union, with a per capita income 40 percent above the Union average. However, its economic performance would have been better if Estonia's economy had not been constrained by the limitations of the Soviet system.[29]

Post-Soviet Estonia: The Age of Independence (1991)

The Republic of Estonia formally declared independence from the Soviet Union on August 20, 1991, during the two-day attempted coup in Moscow. The international community, including the Russian Federation and the United States, recognized the country quickly. The European Union, in turn, recognized an independent Estonia on August 27, 1991. Estonia joined the United Nations on September 17, 1991, and went on to join the World Trade Organization in 1999. Now a free nation, the nascent republic quickly went to work fixing the political and economic mess that the Soviets had left behind.

When Estonia finally became free, it set out to gather scholars, politicians, and academics to draft a fresh constitution, and not base the new political system on what had already been in place under the Soviet occupation, nor the constitution that had been drafted in 1938, during the first period of independence. The new government, a parliamentary system with a prime minister and a president as the head of state, was approved by popular referendum on June 28, 1992.[30]

Under this new constitution, foreigners were allowed to obtain land in Estonia. Mart Laar, who served as prime minister from 1992 to 1994 and from

1999 to 2002, recalled that one of the easiest ways for former Soviet republics to acquire foreign investment was through the sale of property, something explicitly forbidden during the period of Soviet rule. By breaking so rapidly with its Soviet past, Estonia hoped to reassert its independence and to secure a place in Europe. "Your government must be lean and effective and you must have the [capability] to make a clear cut with the past," Laar says. "It's not possible to put the new wine in the old sacks, as it says in the Bible. You'll get a very bad result."[31]

While these political changes were under way, a rapid economic transformation was taking place to bring down high inflation and resuscitate the depressed economy. State subsidies were phased out, while a new currency, the Estonian kroon, was introduced in June 1992, and was pegged to the German deutschmark, one of the most historically stable and inflation-free currencies in Europe. That decision, combined with privatization and a constitutionally mandated balanced federal budget, brought inflation down from 1,000 percent in 1992 to 90 percent in 1993, and eventually all the way to 6.5 percent in 1998. Inflation then rose to an estimated 10 percent, but dropped to an estimated 3 percent by 2010 and 2011, according to the IMF.[32]

Furthermore, under the Laar government, in 1994 Estonia implemented the first flat income tax in the world, at 26 percent—and no business taxes at all. (The flat tax has subsequently been lowered to 22 percent.) Laar has famously claimed that the only book he read on economics prior to taking office was the American economist Milton Friedman's *Free to Choose*, which argues for economically liberal and libertarian policies. He also erroneously thought at the time that a flat tax was commonplace in Europe and the United States.[33] His aggressive and unorthodox economic policies led to the explosion of Estonian economic activity.

As Laar noted:

Attitudes changed surprisingly fast. Thousands and thousands of new small and medium-size enterprises, restaurants, hotels, and shops were established. In 1992, Estonia had about 2,000 enterprises. By the end of 1994, the figure had ballooned to 70,000. Estonia had changed from the country of the working class to a country of entrepreneurs. The incentives to take charge of their own future helped Estonians to avoid massive unemployment.[34]

In 1994, the Laar government also established unilateral free trade on all goods (including agricultural products), which led to rapid technology transfer and high levels of foreign direct investment. By 1996, the World Bank observed that Estonia's rapid growth led to a 500 percent increase in the value of exports to western markets between 1992 and 1995. During that same period, foreign direct investment nearly quadrupled to $209 million, and the average real monthly wage increased by over 600 percent, from $41 to $260. Investments went into real estate development, hotels, retail outlets, supermarkets, shopping malls, and the like.[35]

Beyond legislative and economic reforms, Estonia undertook equally radical changes in its telecommunications strategy. At the time of independence, Estonia's telecom infrastructure was massively outdated, and all calls and data transfers had to be routed through Moscow. On November 28, 1991, the newly created Transport Ministry issued a license to the National Institute of Chemical Physics and Biophysics of the Estonian Academy of Sciences to create a new public data network.[36]

Within four months, a satellite-based connection to Sweden was constructed, and was funded by George Soros's Open Estonia Foundation. Satellite dishes were installed in Tallinn and Tartu; their combined international data capacity was 64 kilobits per second, on par with individual 56 k modems that were popular for individual homes in the United States during the mid-1990s. There were only forty (mainly academic) users on this network the first year.

But what began initially as a small-scale project grew quickly, as foreign companies from Scandinavia began playing an increasingly large role in Estonia's development. By the fall of 1992, the government created Eesti Telekom (Estonia Telecom), with Finland's Sonera and Sweden's Telia each owning a 24.5 percent stake in the new company. Eesti Telekom formed a partnership with Helsinki, and connected to the Internet via an undersea cable to Finland, rendering the previous Swedish satellite system obsolete.[37] Later, in 1998, Estonia established a national communications regulatory board, Sideamet, whose job it was to oversee the telecom industry, including fixed-line, Internet, and mobile services.

In short, the 1990s served as a time of rapid economic expansion and development in Estonia. The ambitious free market approach undertaken by the youthful Laar administration—he was thirty-two when elected prime minister, and his defense minister was twenty-nine—allowed even a young, small, and resource-strapped country like Estonia to quickly gain access to foreign capital. Its Scandinavian neighbors heavily invested in banking and telecommunications, which, in turn, helped to revive the country. As Laar pointed out later: "I've said that the goal of my first government was to turn Estonia from the east to the west and the goal of my second government was to make this turn irreversible."[38]

Indeed, this turn took Estonia on the road to rapid technological adoption, and for someone like Veljo Haamer, it allowed him to zoom ahead.

Tiger Leap and Haamer's Rise and Fall (1992–1997)

In the final days of the Soviet Union, one of the easiest ways to get hard currency was to sell imported products at high margins. That was the strategy of the

Haamer brothers, Hanno and Veljo. The elder brother, Hanno, began importing computers from Russia for domestic consumption, and later he started to buy directly from Asian manufacturers, assembling them in Tallinn and selling them at a 30 percent margin, for about $2,500. "There was no economic reason to buy computers," he recalled. "Computer costs in these times were ten years of an accountant's salary."[39] Hanno Haamer knew that while most people could not afford such machines, those who could clamored for the devices, and he wanted to make sure that he could profit from such a transaction.

One day in 1991, while Hanno was working at a company called Kungla Dialoog, he met a young programmer named Rainer Nõlvak, who had walked into the shop. The two quickly hit it off and lamented the fact that computers were so difficult and expensive to obtain in a rapidly changing Estonia. As Nõlvak recalled: "It was a time when you had to stand in a line for three hours to get gasoline."[40]

The pair decided to start their own wholesale firm, called MicroLink, which could undercut their competitors by assembling computers onsite in Estonia (as Hanno Haamer had done on his own). They would keep a stock of computer parts that didn't change very much in price, like cases and keyboards, and would order, overnight, parts that had rapidly falling prices, like hard drives and RAM. "Computers are like bananas, they rot quickly—the price decreases dramatically all the time," Nõlvak observed.

Given that Estonian manufacturing labor was inexpensive, MicroLink was able to keep its prices down compared to its competitors in Western Europe and North America. But even though these lively entrepreneurs could make a living, as of 1992 the best hotel in the country had no heating to speak of, and contracts had to be signed in −10°C (14°F) weather. (The new company grew to become the largest IT firm in the Baltics and was later sold to Eesti Telekom for just under $9 million.)[41]

Meanwhile, Veljo Haamer and a friend, Peeter Parvelo, started their own retail company, called Korel AS, on August 23, 1991—just two days after Estonia declared independence. They had met only two years before while students in the cybernetics division of Tartu University, and quickly formed a small company called Gensi, which sold accounting software. Parvelo said later it wasn't even a business, but rather "it was more like students making money."[42]

Korel sold the computers that Hanno and his crew assembled. The company sold printers and other accessories to high-end clientele that were willing to pay significant amounts of money for these products. Parvelo described this period as "[the] beginning of robber capitalism in a young Estonian economy."[43]

By 1995, the Estonian government was well into the first term of President Lennart Meri, and was flush with ambitious new ideas. Among them was what came to be known as the "Tiger Leap" (tiigrihüppe in Estonian) project—at the time, one of the most expensive and elaborate projects in the young republic. This idea originated late one night from the minds of two Estonian cabinet ministers—Jaak Aaviksoo (the current Estonian Minister of Defense, who was then the Minister of Education) and Toomas Hendrik Ilves (the current Estonian president, who was then the outgoing ambassador to the United States and incoming minister of Foreign Affairs). Gathered for a late night tête-à-tête at a guesthouse in Tallinn over a few glasses of whiskey, the men had decided that computer literacy had to be part of basic education in all Estonian schools.[44] The idea, simply, was to wire every school to the Internet and to provide PCs to each classroom, all across the 50,000-square-mile territory, from the remote islands in the west to Lake Peipsi in the east, and from the Gulf of Finland in the north to the Latvian border in the south.

Shortly thereafter, President Meri took a trip to the United States, where he asked Microsoft to provide some assistance toward the goal of "making Estonia the model state of information technology."[45] Estimates at the time suggested that the project would cost nearly 1 billion Estonian kroons, over $90 million at the time—a phenomenal amount by any measure.

This ambitious education project officially got off the ground in late 1996, when the Tiger Leap Foundation was created as a public-private partnership between the Estonian government, the United Nations Development Programme, the Open Estonia Foundation, the European Union Phare program, and many companies, including MicroLink. The Estonian Parliament (*Riigikogu*) kicked in 35.5 million Estonian kroons, or almost $3.2 million, for the initial year, with local governments paying the rest, often a figure in the tens of millions of kroons.[46]

At the time, there were an estimated 20,000 to 30,000 Internet users in Estonia, with an estimated 51 permanent connections per 10,000 people— better than nearly all former Soviet states in Eastern and Central Europe combined—but still not nearly as good as neighboring, Finland, which boasted 630 permanent connections per 10,000 people. That said, the rate of Internet usage growth was tripling every year.[47]

As a way to boost Internet usage, the government took a parallel track to Tiger Leap by building a series of public Internet access points in various areas around the country. The first of these was opened on February 21, 1997, at the National Library in Tallinn. At the initial location, a small group of computers was set up for any and all Estonians to come use the Internet, free of charge. As the Estonian daily *Postimees* reported, President Lennart Meri wrote soon after the public access point's opening:

On the Internet, every Estonian's thoughts and words will matter exactly as much as their worth—they will matter equally with the thoughts and words of Americans, Russians, Germans, and Japanese. . . . So be quick to step onto this bridge that has united continents as neighbors and that will take us all, especially the young, right into the next century. Those who walk faster will reach the next century sooner.[48]

Among the other first public Internet access points (*internetipunkt*) were three on the rural western island of Hiiumaa, including one in the village of Sõru, and another in a small community library in the town of Kärdla. There was also a third access point set up in a small farmhouse in the town of Paope, on the northern coast of the island.

The United Nations Development Programme (UNDP) website describes one of these locations:

> Vaike and Valdo Laid's farmstead on Hiiumaa's north coast is one place that will never be the same. The family farm, a small quadrant shared with a handful of horses, cows and piglets, was one of the first locales transformed into a Net visitor centre. Long a hub of news for local villagers, the Laid farmstead served as a post office during Estonia's brief independence before Soviet annexation in 1940.
>
> So when UNDP officials suggested plugging in a computer in the Laid's old granary storage room, Vaike and Valdo stepped into the digital age with ease. "We didn't know what it all meant," says Vaike, "but when the lines formed at our door, we knew our farm had joined the rest of the world."[49]

Interestingly, one of the earliest innovations that was implemented at the time was something rather low-tech—a road sign. Linnar Viik, one of the masterminds behind Tiger Leap, had the idea to create a traffic sign, made from reflective, industrial-grade metal, as a simple way to alert people to the fact that they were near an *internetipunkt*. Estonia became the first nation in the world to use road signs in such a way.[50]

At the time, Minster of Education Jaak Aaviksoo said that "[E]stablishing Internet stations in all the 254 municipalities of Estonia is just as important as joining the EU or NATO."[51]

WHILE THE REST OF THE COUNTRY was beginning to leap, Haamer hit a speed bump. Parvelo felt that Haamer had gone astray by investing some of his profits in a Tartu pub.[52] Meanwhile, Haamer and his wife, Liivi Kruus, were growing apart. Each had a significant other outside of the relationship, and while they both tacitly accepted the situation, eventually their marriage could not withstand these differences. Kruus moved out in January 1997.

Haamer and Parvelo's computer company, Korel, also wasn't doing so well anymore. There were now several businesses competing for the same customers and Korel had lost its market dominance. According to Haamer, the company was on the verge of bankruptcy by early 1997. But Parvelo had a way to save both of them. They would let Korel fail and would start a new company, Korel Systems (Korel Süsteemid AS), which would sell financial software and complete "systems" to large companies, instead of selling small, personal computers.

Parvelo theorized that the hardware market was becoming too tight to be profitable. The pair had already made contracts to sell financial software to large clients, including Tartu University, Estonian National Railways, and MicroLink. Each deal was worth 100,000 Estonian kroons for the license of the software, and installation and other consulting fees would yield ten times that amount for each individual account. Each of these deals was worth significantly more than what Parvelo and Haamer had originally been making in a single year. The money began to roll in with their new business. Haamer had never been richer, and he celebrated by taking a month-long vacation in Thailand and Southeast Asia in the early spring of 1997.

ON SEPTEMBER 13, 1997, Haamer decided to take his new girlfriend out for a night on the town, where well-kept trees line the wide boulevards near the old railway station. They went to a Tartu discotheque that they hadn't been to before. It was a warm summer evening, and they were enjoying themselves, locked constantly in a lovers' gaze. The couple first went to the bar and had a few drinks. Then Haamer took his date by the hand and led her to the dance floor. There, they grooved and gyrated to 1980s American pop music, like Blondie and Talking Heads—Haamer's favorite music.

After an hour, they returned to the bar for some refreshment. But at the bar they saw his girlfriend's ex-husband. The man was clearly drunk, and stammered and staggered over to Haamer to confront him.

"Why were you with my lady?" the man demanded.

"Where is your lady?" Haamer coyly replied.

The man took hold of Haamer's shirt. He was fuming, and within a few moments they were outside to settle their differences. The man was a bit burly and weighed thirty pounds more than Haamer, and looked like he'd had more street-fighting experience. Haamer had been in a few fights during his six-month tenure in the Soviet Army, but he wasn't used to challenging other men to street fights, particularly not ones bigger than him.

Haamer tried to throw a few punches but missed, and a moment later the man laid a heavy blow to Haamer's right temple. Haamer's knees gave way as the dusk-filled sky began to spin around him like a pinwheel. He collapsed to the ground and the man kicked Haamer several times in the head before storming off.

Upon hearing the news of his brother's injury, Hanno jumped in his car and drove the nearly two hours to Tartu. Upon arriving at the hospital, the doctors told him that Haamer had massive head trauma, and possibly serious brain damage. Worst of all, he was in a coma. The doctors told Hanno Haamer that his brother only had a 10 percent chance of being able to function normally on his own.[53]

Haamer woke up in the hospital a week later with his brother and their father, Vaino, at his side.

WHEN THE SYMPATHY had subsided and his visitors went back home, Veljo Haamer still lay injured. He spent months recovering in the hospital, taking on physical therapy and coming to terms with the fact that his short-term memory would be forever damaged. His body had been crushed and his spirit had been shattered. His wife had left him and now his new girlfriend had done the same. What did it say about him that he couldn't stand up to another man?

Further, while Haamer was trying to recover, the business that had been thriving just months before was collapsing around him. Korel Systems was unable to fulfill the contracts that it had already signed to, and some of the partners liquidated the company to start their own firm, leaving Haamer out in the cold. To this day, Haamer argues that he was tricked and lost hundreds of thousands of kroons, potentially millions, which were rightfully his.

Peeter Parvelo, for his part, says that Hanno Haamer represented Veljo's interests following the bar fight and subsequent coma. Parvelo maintains that these business dealings resulted in his losing money as well, and ultimately were not his fault. Further, he says that he is saddened by the loss of Veljo Haamer as a business partner, and as a friend.

As he wrote in an e-mail on June 21, 2008:

> The most painful in all this dirty case is the emotional part. The betrayal of your friends you have had since the university, being friends with families, children, wives, having common parties, interesting trips and journeys, common interests, hobbies. The main lesson I have learned— never do business with your friends if you do not want to get rid of them, money is the dirty thing that does not suit into friendship, whether you have it or not it will lead into quarrel.[54]

After he got out of the hospital, Haamer used the money that he had saved from his previous business to pay his rent and to get by on a day-to-day basis, but most of his time was spent wallowing in beer and sorrow. He spent close to a year in aimless lassitude, seemingly oblivious to his family and friends. His funds began to dry up; he spent his only money on buying one meal at a pub per day—he still has never learned to cook—and didn't drive his car, opting instead to travel by bus. His stubble and unkempt hair made him look older than he actually was at the age of thirty. His belly always rumbled.[55]

This entire period, between the bar fight, the coma, the final breakup with his wife, and the business dispute, nearly destroyed him entirely. If Haamer had not found something to inspire him, to give him a sense of purpose, and to plug him into the society that was taking off around him, he likely would not have amounted to anything.

Tiger Leap II (1999–2000)

While Haamer was floundering, unemployed and depressed, Mart Laar was reelected to a second term as the prime minister of Estonia in March 1999. Laar campaigned on leading Estonia out of an economic slump and marching toward the European Union. By the end of 1999, Laar had another, equally surprising idea as to how to conduct the business of government: the cabinet ministers should meet virtually via an electronic Internet-based system. This plan became one of three major pillars of what some called "Tiger Leap II," a massive technological reformation of Estonian government in the early twenty-first century.[56]

In Estonia, the heads of each ministry—defense, education, transportation, etc.—hold weekly meetings to discuss pending legislation. One minister will propose a bill, the ministers discuss it, and, if approved, it is sent to the Parliament for ratification. Each bill is a lengthy document, sometimes a few hundred pages, which meant that each time a revision was made, the entire bill would have to be printed out, and carried by each minister and their staff members to the meeting. "We're preparing these documents, electronically, we print them out, we make changes electronically and we print them again and sign them—What a waste!" remarked Laar in a later interview.[57]

By March 2000, Linnar Viik made a formal proposal to the prime minister of creating a paperless cabinet meeting. The idea was initially met with skepticism—there were no other countries in the world that Viik could point to as examples—not to mention the cost of such a project. But within a month, the project was approved, and given a 1 million-kroon budget (about $100,000). In less than six months, the system was in place. The ministers got it right away. "We originally thought that the training would take a couple of hours per minister [but after] 20 minutes per minister and they were using it," recalled Viik.[58]

The same building that houses the prime minister's office, a short walk up the cobblestone streets of old town Tallinn from the Parliament, also hosts the cabinet meetings. The e-government system consists of three tables put together in the shape of a U with nothing atop them but LCD monitors, keyboards, and mice at each position. Each PC has a stripped-down Linux-based system that is set to display the online version of the e-government system, enabling each minister to securely log in to view bills. The cabinet quickly agreed that if a bill was not discussed online prior to the in-person meeting, it would automatically pass, and thus the cabinet could move on to more pressing issues. The project saved the government nearly $200,000 annually, and more than paid for itself within the first year. "Cabinet meetings used to take between four and twelve hours," said Tex Vertmann, the prime minister's chief technology adviser, in a 2003 interview with the Associated Press. "Today, they take between ten minutes and an hour."[59]

With the new e-cabinet system set up—drawing visitors ranging from Prince Charles to Elton John—the government turned its attention to another equally ambitious project: digital identification cards. Since 1994, the idea of creating an Estonian electronic identity card had been kicking around the Institute of Cybernetics. By 1997, Tarvi Martens, a software engineer at the institute and a classmate of Haamer's, spearheaded an initiative to conduct a detailed study of what an electronic ID card would look like, and what features it would need.[60]

That same year, the parliamentary ministers began preparing the Digital Signatures Act, and by 2000 two laws passed that became the pillars of Estonia's electronic ID card policy: the Identity Documents Act (1999) and the Digital Signatures Act (2000). These two laws established that identity documents in the form of an ID card would be compulsory for every Estonian and every resident alien, and that each ID should contain a unique set of electronic certificates that would allow the ID card to be used as a digital signature, cementing the use of these cards in modern Estonian life. The Tallinn Administrative District Court ruled in 2003 that digitally signed documents must be legally considered as equivalent to hand-signed documents.[61]

The card, which looks like an ordinary credit card-sized identity card, contains an embedded microchip, similar to many European credit cards. When the card is inserted into a reader, attached to a computer, government officials ranging from ticket controllers on a train to police officers can verify the cardholder's status. For instance, Estonians who ride the Tallinn street tram can buy virtual "tickets" online. When a tram controller comes around to check that people have paid their fare, all she has to do is insert the ID card into a handheld card reader to verify that, indeed, that person has paid. Similarly, if an Estonian police officer pulls someone over, she can instantly verify the validity of that driver's license and insurance through the use of another reader.

Ordinary citizens can get readers for their home computers as well, for under $9 from nearly any bank or supermarket, and then can securely log in to the country's citizen's portal. This portal allows citizens (and legal residents) to interact with various government bureaucracies. In Estonia, with a few keystrokes and a mouse click or two, one can vote, receive speeding warnings, and obtain pension funds. By 2006, nearly 80 percent of Estonian residents paid their taxes online—a number that rose to 92 percent by 2010. By 2008, over 980,000 ID cards had been issued (out of a population of 1.3 million). It is clear that the Estonian ID card and e-government system represent the most ambitious and well-executed project of its kind in the entire world.[62]

Due to the underlying infrastructure of the digital ID cards, Estonia was able to successfully conduct the world's first nationwide Internet-based voting system in 2007. Two years before, Estonia had conducted a limited test for local elections, but no other country has ever attempted a nationwide online election before. Some countries have attempted limited Internet-based voting (mainly in Europe, but also for American military serving overseas, and in the state of Hawaii for local elections), but by all accounts the 2007 Estonian election was a resounding success, with 3.5 percent of the electorate, or more than 30,000 citizens, casting their ballots online, up from the 2 percent in local testing from 2006.[63]

Elections in the United States are astonishingly complicated, with different elections boards in different parts of the country, each with their own set of rules. This plethora of standards is one of the main reasons for the election debacle of 2000. Over a decade later, the United States perennially struggles with electronic voting machines, which have been certified and decertified in many states. Meanwhile, in that time, Estonia has managed to surge ahead.

Granted, Estonia's system would not be possible in the United States, as its system of national digital identity cards would send chills down the collective spine of American civil libertarians. That said, the most succinct and compelling argument for Internet voting came from Tarvi Martens, both the project manager of the e-voting initiative and the head of the ID card project. He told *Wired News* in 2007: "You trust your money with the Internet, and you won't trust your vote? I don't think so."[64]

The Rise of WiFi.ee (2001–2002)

At first, it was just a box with a funny brand name, Orinoco. At PC Superstore, there were lots of new gizmos that came and went. It sat in the mailroom for a few days before Haamer even noticed it. It was called a "wireless access point." The idea was that somehow an Internet signal would be converted into radio waves and then could be received by a laptop equipped with a wireless card. But in Estonia, the access point itself cost 5,000 kroons ($500), and the card another 1,000 kroons ($100)—for Haamer, that was one month's salary.

In 2000, Haamer moved from Tartu to Tallinn to escape the bad memories of the city, and to create a new life for himself. He'd taken a job as a salesman and technician with a company called Baltic Computer Systems. He would travel around town, selling large companies on significant orders of various types of office gear, ranging from PCs to laser printers. When something went wrong, Haamer was one of a few guys who would be sent out in the field. Early the following year, he switched jobs to work at PC Superstore, a small, independent retailer of computers and accessories.

PC Superstore was constantly receiving demo items of various new products on the market. One day while at work in the summer of 2001, the store received a "wireless access point" that promised to provide Internet connectivity wirelessly. Haamer thought that this device was just another fad. Given that the access point alone would cost him a month's salary, he'd never be able to buy one anytime soon.

"It was a toy for rich people," he recalled.[65]

Furthermore, why would he want something that he could get ordinarily, through a standard wired network, at significantly less cost?

EARLIER THAT YEAR, WiFi, began to hit the mainstream. The *New York Times* had written its first story on the subject in February 2001:

> It's coming soon to an espresso bar near you.
>
> And to the rival coffee shop across the street. Not to mention the coin-operated laundry on the corner, the hotel on the next block and the railroad station across town. The local airport may already have it.

The article continued, noting:

> But most access points are and will be commercial, run by companies that will charge for the services—anywhere from a few dollars for a single session to $50 or more per month for unlimited use of the system.[66]

Haamer took the demo model and set it up in the store, where he could test the device without paying a single kroon for the privilege. He quickly and easily installed the software for the wireless card on his laptop. Day after day, he could see the advantage in the store—he was able to use the Internet, completely wirelessly, from the front counter to the stockroom. The WiFi card was a curiosity, mainly. Sure, it was a neat gadget, but given that very few Estonians could afford one, and that they would only be used in high-end offices, what was the point? Haamer's manager told him to send the device back, sure that this overpriced item wouldn't sell.

Haamer kept trying to figure the WiFi gear out, and began looking on the Internet for more information about WiFi. He quickly became an avid reader of wifinetnews.com, the first blog devoted to WiFi, authored by Glenn Fleishman, the same journalist who wrote the February 2001 piece for the *New York Times*. Haamer learned about community networking projects in Seattle, San Francisco, and New York. That spring, the first Estonian public WiFi access points became available, and were installed by the local Internet service providers Eesti Telefon (now Elion) and Uninet.

By March 2002, Haamer and a small group of dedicated WiFi activists decided to found WiFi.ee, a nonprofit organization, whose mission is to "make the Internet easily accessible throughout the country, particularly at places which are attractive to tourists and those that are close to crossroads. Our ongoing efforts include regular inspection of the capacities and compatibility of new wireless Internet areas."[67]

A few months later, PC Superstore went out of business—but Haamer kept the WiFi gear. The price—less an employee discount—was taken out of his paycheck.

Out of a job, Haamer spent most of his free time online. He learned about various types of devices, gateways, routers, and access points. He absorbed the specs of directional antennas while half-heartedly looking for employment. WiFi was the cool new thing, but until the cost of an access point dropped under $100 and laptops dropped under $2,000 in Estonia, there was no way that ordinary people were going to be able to use it.

As the spring continued, Haamer set to work on getting the first free WiFi hotspots up and running in the country. He set his sights on a hotel café that he frequented for morning tea and breakfast—Café Mademoiselle, inside the Grand Hotel Tallinn at the bottom of Hermann Hill, the site of the 1989 raising of the Estonian tricolor. The manager of the café was very skeptical about WiFi at first, both because the gear was expensive and because there was hardly anyone—Estonians or foreign tourists—who came to the café with their laptops on a regular basis. Haamer argued that soon free WiFi would be as common-place (and perhaps one day as cheap) as newspapers hanging on a rack for café-goers to read. He posited that just as there was once a day when patrons would come to a café with free newspapers, the same would be true with WiFi. If Café Mademoiselle didn't do it first, their competitors would.

After some cajoling, the manager acquiesced and agreed to spend the hundreds of dollars on the WiFi gear, plus a couple hundred more to have Haamer come and install the wireless network in the café. Haamer was thrilled. Day after day, he would return to the café and enjoy his morning tea with a side of WiFi. On the coldest of mornings, he lingered inside just a little longer to relish the steam filling his lungs and the e-mails filling his inbox. After a few days, he realized that there needed to be a quick but simple way to alert passersby that this café was WiFi-equipped. Short of entering the café and opening one's laptop, it was impossible to know if there was a network or not.

Having seen the example of the Internet road signs from Tiger Leap, he came to an easy conclusion: WiFi.ee needed road signs. The signs, which cost about $50 each to make, are industrial-grade, reflective road signs. They proudly state: "wifi.ee / Tradiitaa Interneti leviala / Area of Wireless Internet."

Haamer quickly went after a few of his favorite businesses in Tallinn and Tartu, including the Wilde Pub in Tartu, a large Irish-themed watering hole. The pub received the first WiFi.ee sign in the country, bolted to the exterior brick wall and easily visible to passersby. Soon after, WiFi networks (and appropriate signs) popped up, thanks to the seed planted by Haamer, in the Port of Tallinn and the Tallinn International Airport. In some locations, like smaller cafés, he created a smaller version of the logo on a sticker—exactly like a credit card sticker in the window of a restaurant. Stickers were cheaper to manufacture than the signs, yet equally effective at spreading the message of WiFi and alerting the public to its use.

At the time, Haamer wrote about the origins of the signage:

American experience tells us that a network alone is not enough. Truly widespread use of the wireless Internet requires usage areas that are easy to find and to identify. Information about available areas and conditions must be easily found by local residents and by tourists. For tourists, the biggest challenge in the United States is precisely this lack of information. There has been no unified source of information about all categories of providers.

WiFi.ee aims to be the Internet place where one can find all WiFi-related information. All open WiFi areas in Estonia, moreover, are marked with the WiFi.ee sign, irrespective of whether they provide free access or not. Bars and shopping centers have stickers similar to those from credit card companies. Open areas are marked with a WiFi.ee traffic sign.[68]

Precisely because Haamer set an early precedent and was able to convince business owners to provide free WiFi, such access today is more prevalent and is less expensive than anywhere else in Europe, and possibly anywhere else in the entire world. Just as Haamer was building up his wireless empire, a new generation of Estonian technology companies was being born. Estonia was booming faster than ever before—and perhaps the most obvious sign that Estonia had finally made it on its own was that the U.S. Peace Corps office in Tallinn closed its doors on June 17, 2002.[69]

Ascent of the Estonian Tech Sector (2003–2007)

Institute of Cybernetics

Tallinn

April 2003

"Hey guys, it looks like it works now," said Taavet Hinrikus, giddily holding a laptop to his ear, like an oversized cell phone.

He was standing at one end of the cavernous hallway, deep in the bowels of the Institute of Cybernetics, where Skype had set up shop earlier that year. A few of the other employees were standing nearby, at various lengths down the hallway, looking equally ridiculous with computers opened to their ears and mouths. The rest of the employees watched, their glances darting from one end of the hallway to the other. The simple idea behind Skype was to make calling over the Internet as simple as picking up a phone—and those calls, to anywhere in the world, should be free. Hinrikus was one of Skype's original ten employees, and was one of its first managers at the age of twenty-two.

Toivo Annus, another manager, held up his laptop at the other end of the hallway. "Yes," he spoke, with the usual reserved modesty of most Estonians. "I can hear you."

This experiment proved that Skype worked—one could now easily make calls from any computer to any computer via the Internet, regardless of the distance between the two machines. A few months later, on Friday, August 29, 2003, the team released the first version (for Windows) of Skype. On the following Monday there were 10,000 downloads even though the company had done no marketing or any kind of public announcement. By the end of the week, 60,000 users had downloaded the software.[70]

Skype wasn't the first company to conceive of using the Internet to make phone calls, but it quickly became the simplest, easiest, and best. Prior to Skype, Internet phone calls were unreliable at best, and essentially only for those who were savvy and patient enough to want to try it. Getting such calls to work before the days of Skype required proper configurations of firewalls, network latency, and audio settings, or, as Hinrikus, Skype's former director of strategy, recalled later: "If you were a geek and you took the effort to configure it, you could make [those pre-Skype applications] work."[71]

He added, underscoring Skype's intention of simplicity: "Skype was designed for your mom to call you, or for you to call your mom from college." Hinrikus and his then-girlfriend became two of the original beta-testers, as she was studying in the Netherlands at the time. As they talked for hours, for free, each night, he became very motivated, very quickly, to make sure that the software worked properly.

Two Scandinavian entrepreneurs, Niklas Zennström and Janus Friis, conceived of this new company as a telecommunications force to be reckoned with. In fact, the story of the company name reveals a little bit about the early history of the company itself. One of the early names that Zennström, Friis, and the other founders considered was: "Sky peer-to-peer," which seemed far too long and cumbersome. That then got abbreviated to Skyper. However, there already were Skyper domain names. So, wanting something unique, they shortened it further, to Skype.[72] The company's founders had been the previous masterminds of Kazaa, a controversial peer-to-peer application—a more advanced version of the original Napster—that allowed users to easily pirate music, movies, and software. And just as Kazaa paved the way for upending the music and film industry, spurring the major studios to get behind legal digital downloads, so too did they want Skype to revolutionize the telecommunications industry. Zennström and Friis had the idea of combining the underlying technology of Kazaa—which allowed for a decentralized network, with each computer sharing some of the load—with making calls over the Internet.

As Friis told CNET News.com in 2003: "After Niklas Zennström and I did Kazaa, we looked at other areas where we could use our experience and where P2P technology could have a major disruptive impact. The telephony market is characterized both by what we think is rip-off pricing and a reliance on heavily centralized infrastructure. We just couldn't resist the opportunity to help shake this up a bit."[73]

The duo quickly hired an Estonian, Toivo Annus, who managed three core programmers from Kazaa, Jaan Tallinn, Priit Kasesalu, and Ahti Heinla. This trio—possibly the best programming team in all of Estonia—had known each other since 1992, when they founded Estonia's first video game company, BlueMoon Software. The core developer team was set up in Tallinn, mainly because there were a lot of high-quality programmers available for relatively cheap. Like many other projects in Estonia, Skype moved fast, going from concept to launch in eight months.

In late 2005, eBay acquired Skype for $2.6 billion, then sold the company to a group of private investors for $2.75 billion four years later. The team of original programmers, including Hinrikus, went on to found a venture capital firm called Ambient Sound Investments as a way to find, and profit from, the next generation of Estonian innovation. (But perhaps Skype finally went mainstream when Prime Minister Andrus Ansip gave President George W. Bush a Skype phone during the American leader's visit to Estonia in November 2006.)[74]

Some of that next generation may come from a part of Tallinn that is undergoing a similar massive transformation. Old town Tallinn is lined with cobblestone streets and bars filled with European tourists that stay open well into the sunny summer evenings. In Vanalinn, the old city, a half-liter beer costs nearly double (about $6) what that same beer costs almost anywhere else in the country. This small medieval core—one can walk across it in around twenty minutes—is surrounded by a ring of more modern-looking buildings, including cinemas, high-end apartments, luxury hotels, and fancy restaurants that sit on heated balconies overlooking the rest of the city. A few kilometers toward the south is the central bus station and the airport, and just next-door, Ülemiste City, Estonia's new tech incubator.

Upon a cursory glance of the 81.5 acres (33 hectares) of industrial workspace on the edge of town, the modern successes of downtown Tallinn seem not to have reached this place. Large brick warehouses, with stacks of old tires around them, dominate the scene. Dust and sand blow between the stodgy structures. One white brick building still has red bricks that spell out in the apex of the gable the Russian words "Communist Party of the Soviet Union Forever!"[75]

The entire facility, previously known as Dvigatel, was originally designed in the late nineteenth century to manufacture railroad cars. After World War II, it soon became a factory building components for nuclear reactors, submarines, and other top-secret military equipment. At one time, it was one of the largest military-industrial facilities in the entire Soviet Union. One of its multi-story workshops was even designed with the windows angled and shaded so that a view inside was impossible from the airport—several hundred yards away. In 1996, the entire facility was privatized by a group of entrepreneurs who hoped to continue the work that had been done there for decades. But the orders from Russia, not surprisingly, ceased, and the expected business from Western Europe never materialized—not then, and not for the next decade.[76]

By 2005, Tallinn's boom was well under way—and just like in Silicon Valley, land values skyrocketed. A new entrepreneur, Gunnar Kobin (a former CEO of the Estonian branch of the Finnish grocery giant Kesko Food), was brought in to rethink the use of entire facility. He quickly decided to scrap the entire existing infrastructure, knock down most of the buildings, and plan an entirely new high-tech business incubator center, spending $150 million on the development of the space.[77] The plan was to create more than 2 million square feet of new office space, hundreds of thousands of square feet of new retail space, and a new convention center for the city. The company would partner with the city to provide incentives for Estonia's largest tech companies to relocate here, and then use their presence to attract smaller retailers and even restaurants. Plans called for a new six hundred-person restaurant, which would make it the largest in the country. By April 2008, interest had grown so much that office space rental prices had jumped from 180 kroons per square meter ($18 per ten square feet) to 250 kroons per square meter ($25 per ten square feet) in one particular building—still on par with the rest of Tallinn.

The idea is to turn this relic of Soviet-occupied Estonia into a dynamic business incubator, more specifically, something along the lines of a university, Silicon Valley, or its most well known example, Google.

As Kobin stated in a 2008 press release: "Our goal is to develop Ülemiste City continuously so it would be a place where you love to work. The Google campus Googleplex is a good example for us. It's been built so that people would feel good there."[78]

By 2008, some of Estonia's largest companies, including Webmedia and MicroLink, had relocated there, as had a business school, a few government offices, and even the local offices of large multi-national corporations, like Nestlé and Johnson & Johnson. Large parts of Ülemiste City are currently operational, and Kobin expects it to be fully completed by 2016.

Consequences of Heavy Online Reliance: Estonia's Cyberattack (2007)

It wasn't an accident that the first large-scale cross-border politically oriented cyberattack was launched against Estonia. Like any good attack, all it needed was a spark. It began with a statue.

Until the early dawn of Thursday, April 26, 2007, a larger-than-life dark bronze statue of a caped Soviet soldier looking somberly toward the ground had stood largely undisturbed in a Tallinn city park since 1947. The monument commemorates the USSR's war dead in defeating the Nazis in the Baltics. To the ethnic Russians who never left Estonia at the end of the Cold War, the statue symbolizes the bravery of their countrymen and the liberation of the Soviet Union. To Estonians, the statue is yet another painful reminder of the vicious history that has long-divided these two groups. There had been talk of removing the statue in previous years, but with vague threats from the Russian Federation that moving the statue would cause trouble for tiny Estonia, the government had resisted for sixteen years.

At 4:30 A.M., following a vote by the Estonian Parliament earlier in the year, the initial process of removing the statue began. Estonian riot police encircled the statue while a two-meter-high fence—the first step in investigating the possibility of exhuming bodies when the monument itself was moved. Protestors—mainly ethnic Russians—immediately swarmed the statue, chanting at the assembled officers: "Shame on Estonia!"[79] By nightfall, more than a thousand people had gathered, spilling out from the triangular city park onto side streets. Most of the protestors strongly felt that the possibility of moving the statue was an affront to Russian pride and history.

The crowd turned violent and began smashing adjacent shops and storefront windows. By the end of the night, one person had been killed, dozens injured, and hundreds arrested. An emergency government commission convened early Friday morning and decided to hastily move the statue to an undisclosed location that same day. This tactic spurred a second day of more muted protests.[80] Other, smaller protests took place in smaller cities in the northeastern part of the country, near the Russian border. Oddly, no violence was reported in the border town of Narva, which has the largest number of ethnic Russians in Estonia.

Unknown to most people, save Hillar Aarelaid, the head of Estonia's Computer Emergency Response Team (CERT), and a handful of others, the worst was not the street riots, but what was to come on the Internet. As he told the *New York Times* almost a month later: "If there are fights on the street, there are going to be fights on the Internet."[81] He had already called in extra staff and brought in extra computer hardware to deal with the anticipated onslaught that government computers might face.

By the time the first assaults began on Friday evening—unprecedented amounts of traffic began pummeling Estonian web servers—Aarelaid was in Ireland attending a cybersecurity conference. That evening, while he was at a pub with some of his colleagues, he got a call from Margus Kreinin, the head of the Infrastructure Department at the Estonian Informatics Centre (RIA).

"It's started," Kreinin said. This was what they had been waiting for.

The pair chatted for a few minutes, and they hung up—there wasn't much that Aarelaid could do from Ireland, so he went back to his beer. He would return to Tallinn the following morning. This attack was the beginning of what came to be known in Estonian government circles as "Phase I," which was characterized by seemingly innocent cases of "hacktivism" perpetrated against government sites. They were, at best, meaningless pranks.

On Saturday, April 29, 2007, life in Estonia appeared to have returned to normal. The damage from the riots had been cleaned up and the weekend was beginning. Most people were looking forward to Volbriöö, the spring festival eve holiday on April 30 that involves general revelry and overindulgence, particularly among the college students. May 1, or Kaatripäev (Hangover Day), is viewed as the first real day of spring, when the months of long nights and frozen temperatures are finally complete.

As the sun rose, a new wave of assaults began. Thousands upon thousands of spam e-mail messages were automatically sent simultaneously to the mail server of the Estonian Parliament, taking it down fairly quickly. One of the major newspapers, *Postimees*, was similarly assaulted. As the attacks began, ordinary citizens were not terribly affected, other than perhaps being unable to access banking, government, and media websites. But the symbolism and scale of the attack shook Estonian business leaders and the government into action.

From his home office in Rakvere, in northeastern Estonia, near the Russian border, Ago Väärsi, the newspaper's IT manager, watched in terror as the available bandwidth dropped further and further. Normally, the paper only used 20 to 30 percent of its bandwidth—on a particularly heavy day, maybe 70 or 80 percent.[82] But this time, the newspaper had gotten 2.3 million page views in a single day—nearly double the entire population of Estonia and more than the total number of Estonian speakers in the world.

In this type of attack, known as a "denial of service" attack, millions of computers around the world—previously infected with a piece of malicious software—are activated and collectively become a "botnet." This botnet suddenly bombards a particular target computer with an overwhelming amount of data, making it completely unusable. This tactic is analogous to suddenly flooding a moving highway with thousands of cars. Traffic gets slower and slower and ultimately will stop entirely. This is what happened to various sites on the Estonian Internet. The bots are often installed and executed without the individual computer users realizing what has gone wrong, and are controlled by a handful of people who are behind the master computers that are sending out the commands to the legions of bots.

Across town, other sites were taking heavy hits, including political websites, such as those of the Reform Party and of Prime Minister Andrus Ansip. After all, his party had organized the statue's removal. The website of the president, Toomas Hendrik Ilves, was also quickly overwhelmed. Bank websites were hit hard as well, and at one point a major ISP, Elisa, went down entirely. In total, more than two hundred websites were targeted. Despite the cyber-damage, many Estonians likely were either unaware of the attacks or didn't seem to care. According to a report in *Postimees* at the time, nearly half of those surveyed were not affected at all.[83]

There were even a few examples of more insidious attacks. At one point, the hackers went so far as to post a fake letter of apology from the prime minister, purporting to apologize for the removal of the statue. The attacks ceased on

May 18, 2007, at precisely midnight, Moscow time (one hour ahead of Estonia)—exactly two weeks after they had begun. Phase II was over.[84]

SO WHO ORGANIZED the attacks? No one can be completely sure. Many Estonian government officials are quick to blame Russian agitators from outside Estonia, as there was a great deal of explicit instruction explaining how to attack Estonian web servers given on Russian-language online forums. Some evidence, like server logs, show that the attacks may have originated from Russia, and possibly even from the Kremlin itself. As General Johannes Kert, an advisor to the Ministry of Defense, put it, the government cannot directly accuse the Kremlin of being involved: "In such a geo-political place where Estonians are living, we prefer to be politically correct."[85]

But any IT expert, including Aarelaid himself, knows that server logs, IP addresses, traceroutes, and other digital forensic evidence can easily be faked.[86] In the local and international press during and since the attack, the Russian government has firmly and repeatedly denied that it was involved. Still, those denials don't stop Estonian officials from holding on to their hunch that the Russian government was somehow involved, implicitly or explicitly. The historical relationship between Estonia and Russia enables such conspiratorial thinking.

When asked who he thought was behind the attacks, Aarelaid expressed his frustration at the lack of definitive or provable information about the perpetrators:

> It's not about which country actually this IP address comes from. It's much more important to understand which person was given this command to attack. But usually it's quite hard to prove in that level that can be [valid in a court of law]. What we saw was that some people, a lot of people, talked about how to attack Estonia and when to attack Estonia, very publicly. The only common thread was that they talked about it in Russian, in Cyrillic, and in nonhacking forums where they usually do not discuss this kind of thing.
>
> And it suddenly went on that everyone started to scream: "Let's attack Estonia, let's take their service down. Estonia is bad, Estonia is fascist, Estonian prime minister is fascist," etc. But who was behind it? Was it the French government, the Russian government? We don't know. And it's very hard to prove. Yes, it was in Cyrillic. Yes, it was in Russian. Yes, we can say so that there was some kind of orchestration. Yes, there was some kind of coordination.
>
> But people can also say that the coordination was very bad and that there wasn't any coordination at all. But was it intentionally so that it seems uncoordinated? Or was it uncoordinated? We don't have answers

today. Maybe we don't have this answer in ten years. Maybe this answer will come up in some archive in fifty or one hundred years that there was a secret document of one government to do so. We don't know.[87]

It's not only Estonia's top Internet cop who doesn't know what's going on. Margus Kurm, Estonia's chief prosecutor, points out that the nature of cyberattacks—they leave very little physical evidence behind—makes it nearly impossible to go after any suspects and charge them with crimes.[88]

In the wake of the attacks, Kurm and his staff began trying to determine whether they could legally prove that any crimes had been committed. But the only evidence they had were some of their own local sever logs, which provided limited data as to which computers were attacking Estonia. Further, that data was not 100 percent reliable, without confirmation from other similar records outside Estonia. Domestic logs could provide some material if the attacks came from within the country, but without similar logs from Russia or other "non-friendly" nations, the Estonian government had little hope of tracking down the perpetrators.

Before the attacks were over, a nineteen-year-old Estonian citizen and ethnic Russian named Dmitri Galushkevich was detained. He quickly confessed to attacking government computer networks, which is punishable—according to the Estonian Penal Code Section 206, subsection 2—with up to three years in prison.[89] But Galushkevich said that he acted alone, based on instructions that he read online. He didn't have any knowledge about who the masterminds or perpetrators in other countries might be.

In the succeeding months, the Estonian government requested further information from Russian authorities. Estonia had a list of IP addresses that appeared to originate from within Russia, and needed the help of their neighbor to conduct further investigations, and perhaps find new suspects. But the Russian embassy in Tallinn and the Kremlin gave their Estonian counterparts the runaround, arguing that technicalities of the treaty between the two countries prevented Russia from providing this information. Besides, the Russian constitution forbids the extradition of its own citizens, so there was no way for Estonian authorities to interrogate or even depose any Russians. Based on evidence that he's seen, and Moscow's lack of cooperation, Chief Prosecutor Kurm says that he is confident that the leaders of the attacks are in Russia. He does add that "we have no evidence and no information that this was the Russian government."

Still, Kurm holds little hope of ever gaining any further information that could be legally useful for prosecuting anyone for cybercrimes against the Republic of Estonia. In an interview in July 2007, he admitted: "The status is that we haven't got any information from Russia and I'm quite sure that we will not get any information."[90]

On January 25, 2008, Dmitri Galushkevich officially pled guilty to attacking Estonian websites. He had to pay a fine of 17,500 Estonian kroons, or around $1,700, and received only probation—no jail time. The case was closed, and no further legal action was taken against anyone, largely because, in the words of Kurm, "Russia refused to cooperate."[91]

Since the cyberattacks, Galushkevich remains the only person publicly named as having a connection to the events of April and May 2007 in or outside of the country. Estonian authorities say that their hands are tied and that all they can do is encourage further international participation in the future with their NATO and European Union allies.

In September 2007, in an address before the United Nations General Assembly, Estonian president Toomas Hendrik Ilves underscored the need for greater international cooperation concerning cyberdefense:

> If in the past people were connected by sea lanes and trade routes, then today we are ever more connected by the Internet, along with the threats that loom in cyber-space. Cyber attacks are a clear example of contemporary asymmetrical threats to security. They make it possible to paralyse a society, with limited means, and from distance. In the future, cyber attacks may in the hands of criminals or terrorists or terrorist states become a considerably more widespread and dangerous weapon than they are at present.
>
> Cyber attacks are a threat not only to sophisticated information technological systems, but also to a community as a whole. For instance, they could be used to paralyse a city's emergency medical services. The threats posed by cyber warfare have often been underestimated since, fortunately, they have so far not resulted in the loss of any lives. Also, for security reasons, the details of cyber attacks are often not publicised. In addition to concrete technical and legal measures for countering cyber attacks, governments must morally define the cyber violence and crime, which deserve to be generally condemned just like terrorism or the trafficking in human beings. Fighting against cyber warfare is in the interests of us all. This requires both appropriate domestic measures as well as international efforts.
>
> In April and May of this year Estonia successfully coped with an extensive cyber attack, and we are prepared to share with other countries the know-how we have acquired. We call upon the international community to cooperate in legal matters in questions concerning cyber security. But, since this whole subject is a relatively new field, it is essential to establish an appropriate legal space. As a first step, we call upon all countries to accede to the Convention on Cyber Crime of the Council of Europe. The Convention is also open for accession for non-members of the Council of Europe.

We should move ahead and create a truly international framework to combat these vicious acts. The Global Cybersecurity Agenda of the International Telecommunications Union, launched by the Secretary-General in May, is a very important initiative for building international cooperation in this field. Estonia also agrees with the assessment of the specialists of the United Nations Institute for Training and Research, that a globally negotiated and comprehensive Law of Cyber-Space is essential, and that the UN can provide the neutral and legitimate forum for this task.[92]

The best current example of this increased multilateral participation is the opening of the NATO Center for Excellence in Cyberdefense, housed at the Estonian National Defence College Training and Development Centre of Communication and Information Systems, on a military base on the south side of Tallinn.

Although cyberwarfare has gotten a lot of attention in recent years, Estonia has been well aware of potential cyber threats since 2004, when the Center for Excellence was first proposed to NATO. Given Estonia's irreversible integration with online technologies in government, banking, and everyday life, some officials in the government knew that this online infrastructure would, sooner or later, become a military target in some way. However, the proposal languished in bureaucratic limbo until it received newfound attention following the cyberattacks of 2007.

The center was formally opened in May 2008 and houses cyber researchers from the Estonian military, in addition to representatives from various NATO countries. In a particularly Estonian touch, the building also has its own sauna, where apparently many after-hours meetings take place.

Kenneth Geers was the first foreign representative and moved to Estonia in late 2007 from the United States. He describes the center's role as being a think tank for Estonia and for all of NATO—to analyze possible cyberwar strategies, tactics, and responses. In addition, the center is tasked with devising NATO cyberdefense doctrine and determining how best to integrate legal tactics to prosecute future attackers. However, given how new the center is, and given how unprecedented its studies are, it remains unclear how much of their research will be publicly available and how much of it will stay within the higher echelons of NATO bureaucracy. Geers admits that his job is tough, and that studying cyberattacks, cyberterrorism, and cyberwarfare is not easy:

One of the first problems in cyber conflict is knowing even whether you're under attack. A very successful cyberattack would achieve success without you even knowing whether you were under attack. So it could be stealthy espionage via the Net, which is possible today. Just in terms of hacking into computers and stealing the information on it in a way that is stealthy. So you wouldn't even know that your adversary was aware of your plans or your goals.[93]

Once authorities have identified what is and isn't an attack, getting foreign countries to cooperate with one another is difficult. That's why, he says, this level of cooperation within NATO is crucial—to cut down on the number of havens where these perpetrators can hide. He adds:

Hackers know that if they route their attacks specifically, internationally, but more specifically through countries that do not have good international relations, that the chances of good law enforcement in an investigation are extremely low. So if I was an American attacker wanting to hit an American bank, I would route my attack through Russia, China, Sudan—countries that have poor diplomatic relations with the United States or countries that have infrastructure that is not highly advanced.[94]

While no one can be sure exactly when or how the next large-scale cyberattack will happen, it is clear that Estonia will be prepared. Moreover, its role in NATO, the European Union, and the world at large is being forged out of this experience of high-level IT security and cyberdefense.

Estonia in a Post-Cyberattack World (2007–Present)

So how did all of this happen? How did a country that had been occupied for centuries and particularly brutalized by the Soviets emerge into the early twenty-first century as an affluent, educated, and completely wired nation? Why is Estonia the country that has free WiFi on every corner, and the nation that conducted the world's first national election via Internet?

Estonia's history, even its centuries-old history, plays a large part in the country that has emerged today. First, Estonia was a well-educated nation even before it became independent again at the end of the twentieth century. The first Estonian-language books were published in the mid-sixteenth century. Its major university, now called the University of Tartu, was established in the early seventeenth century, and then as now, it held the highest standards for academic achievement. During the end of that century, Sven Dimberg, a Swedish mathematics professor, became one of the first professors in the world to lecture based on Newtonian mathematics and physics.[95]

In the eighteenth century, religious texts and calendars were being printed in the Estonian language and literacy was encouraged and promoted by Protestant missionaries who encouraged study of the Bible in one's native tongue. By the mid-nineteenth century, J. V. Jannsen, who encouraged his countrymen to take pride in their Estonianness—distinct from the dominant German (and sometimes Russian) language and culture of the upper class—founded the first major nationwide newspaper, *Postimees*, in 1857.

Jannsen used his paper as a medium to rile up the populace, encouraging education, private ownership of land, and Estonian pride. In fact, it was Jannsen who first used the printed word "Estonian" (*eestlane*) instead of "countrymen" (*maarahvas*) during the first year of publication.[96] He became one of the key figures in a growing academic, literary, and eventually political movement that came to be known as the National Awakening. The words of his poem "Mu isamaa, mu õnn ja rõõm" (My Country Is My Happiness and Joy) now make up the lyrics of the Estonian national anthem. (His daughter, Lydia Koidula, went on to write equally nationalistic poetry that urged her fellow citizens to also take pride in their language and nation. Her poem "Mu isamaa nad olid matnud" [My Country Is My Love] later became the unofficial national anthem during the years of Soviet occupation.) The tail end of the nineteenth century became an important period for the development of Estonian identity and nationalism and eventually paved the way for the birth of the modern Estonian republic in 1918, just at the beginning stages of the Bolshevik Revolution.[97]

So what does nineteenth- and early twentieth-century literary and political history have to do with technology that didn't arrive in Estonia until the conclusion of the twentieth century? Without this high level of education and political and economic sophistication as a backdrop for Estonia's reemergence onto the world stage, the Internet would not have been able to take root nearly as rapidly as it did.

The simple fact that Estonian literacy was at 96 percent by the beginning of the twentieth century—compared to just 44 percent in Spain—shows that Estonians were well educated and were exploring new ways of thinking. The Estonian scholar and historian Toivo Raun notes that the Estonian society of the late nineteenth and early twentieth centuries "appears to have internalized a positive attitude toward learning, and beginning with the awakening of the 1860s, education and culture were the dominant concerns of the Estonian national movement."[98]

It is an obvious but often overlooked fact that one cannot use the Internet unless one has the ability to read, preferably in one's native language. As such, it is baffling that there are so many well-intentioned nonprofit organizations in the West devoted to bringing computers and Internet access—MIT's One Laptop per Child being the most egregious example—when in fact much of the developing world cannot read or write in their native language.

As such, it is no wonder that Estonia was able to emerge from the Soviet yoke in 1991—as if from a hibernation—and immediately go to work on constructing a viable and prosperous nation, fueled by an ambitious combination of Soviet-instilled scientific education and accelerating nationalism. With education at such a high level to begin with, it is no wonder that the Soviets forbade the study of the dangerous liberal arts. During the Soviet occupation, Estonian scholars both in and outside the nation knew and understood that self-determination and self-reliance would be what would carry Estonia forward. Many nations around the world, such as Senegal, cannot and will not be able to take advantage of the Internet and its benefits until they first solve the problem of educating their citizenry.

Another reason that Estonia was able to succeed so rapidly was its proximity—in both geographic and cultural terms—to its neighbor, Finland. Its neighbor to the north was officially neutral during the Cold War, and thus was not subject to the harsh treatment by the Soviets, as Estonia was. But like Estonia, Finland has been occupied by foreign powers (Sweden, then Russia) for centuries. The story of twentieth-century Finland is one of balancing its survival against an expansionist neighbor, the Soviet Union. Because of this, much of the Finnish political energies have been expended on preserving the sovereignty of the state and projecting that survival into the future.

As Michael Castells and Pekka Himanen, two scholars of Finland, write in *The Information Economy and the Welfare State,* their seminal work on the country:

> The Finns do not feel that their country has come of age, unlike other European countries with thousands of years of history. The information-society project suits a young country that is still partly in search of an identity. With little history to build their identity on, the Finns are oriented to the future. For Finland, the "post-survival" culture is something that is being created now; looking forward and not backward.[99]

This Finnish orientation toward the future is built into the Finnish psyche through a positive disposition toward technology. Finland was one of the first nations in the world to adopt the electric light and the telephone in the late nineteenth century. And yet, as former prime minister and Estonian historian Mart Laar notes: "In 1939 [the year of Estonia's forced annexation], Estonia's living standards and way of life were more or less the same as neighboring Finland's."[100]

Indeed it is no surprise that Finland was a pioneer in the development of cellular phone technology, creating the Nordic Mobile Telephony standard in 1981. By the next year, Finland had the largest mobile phone network and biggest mobile phone market in the world.[101] In 1990, Finland became the first country in the world to have a commercial GSM operator, the nearly universal standard for mobile phone communication today.

As Finland became more and more developed economically, information about Finland began to trickle across the gulf. Throughout the Cold War, Estonians either smuggled in illegal radios and televisions so as to access Finnish media, or figured out how to surreptitiously modify their own Soviet radios to be able to receive Finnish stations. Estonians thus became exposed to Finnish commercial products—including personal computers, and the first generation of mobile phones.

As the Soviet grip on Estonian society began to wane, more and more Estonians began to be exposed to the economic and technological advantages that their cousins to the north had. In 1990, Linnar Viik, one of Estonia's great technological thinkers and visionaries, began his master's degree studies at the Helsinki University of Technology. He continued through 1996 in post-graduate studies in the Department of International Economy and Technology. His exposure to the Finnish spirit of innovation—and specifically to new "advanced" technologies like Internet access and mobile phone usage that were not available to Estonians at the time—significantly helped his countrymen understand the benefits and efficiencies that information and communications technology could bring to Estonia.

"Finland has been a role model for the Nordic welfare society, based on innovation and a strong engineering culture and a strong focus on new communications technologies," he says. "However, the path we took—we realized that we couldn't afford to buy the technologies that Finns had, so we took a number of shortcuts around it."[102]

But while Viik and others were studying how to improve Estonia, ordinary citizens' lives improved in material ways. Even basic communications technologies—like the personal computer—were now available to the masses, when essentially no other writing technology was available to them before. In essence, Estonia to a large degree was able to leapfrog over older legacy technologies—probably more quickly than any other country in the world. As the Estonian writer Andres Langemets noted in 2000:

> When PCs became more prevalent, the inertia of the conservative wing of scholars lasted for years, but Estonian humanities scholars adapted to computers much faster. We had no inertia, because we did not have good established telephone communications, we had no faxes and no electronic typewriters. Estonian society ended up sitting in front of computers at a lightning pace, almost straight from the stone-age, because it had to, and because it was unavoidable. Because it was only yesterday that even the most primitive typewriters could be obtained only with great difficulty and with special permits—as recently as the start of the 1980s the Writers' Union organized many campaigns until permission was given to acquire Yugoslav typewriters.[103]

As Viik returned to Estonia toward the second half of the 1990s, he became very much involved in various government projects, such as Tiger Leap, which invested significant portions of the state budget in Internet access and computers for schoolchildren. In other words, Estonia's nationalistic drive took the nation on a course far away from its Russian past and straight into the arms of Finland, and a free, prosperous Europe. Estonia is very close to becoming even more fully integrated into the European Union, as it plans on adopting the euro in 2011.[104]

Once using computers and the Internet took root (essentially the 1990s), it wasn't long before Estonia began to have its own innovative ideas as to how to improve the foundation that it had built. This is a country that used public websites to publish the names of students who were delinquent in paying back their low-cost government student loans starting in 1999, and used a similar system to publicly shame drunk drivers. In 2010, Estonian judicial officials launched yet another website listing the names of delinquent Estonian fathers who had not complied with court orders to pay child support.[105]

Bringing access to all students was one thing, but refashioning the entire way the government operates and the interactions between the citizenry and the government is something else entirely—and yet, that's exactly what Estonia did, by creating its e-cabinet setup in 2000. That, in turn, begat digital identification cards, which in turn begat e-government, which in turn begat Internet-based voting.

But the most amazing thing about Estonia's technological innovation in both the public and private sectors has been its ability to rapidly export its technological knowledge and prowess. Within two years of the formation of the e-cabinet, Linnar Viik and two colleagues, Arvo Ott and Ivar Tallo, created the e-Government Academy, as a joint initiative with the Government of Estonia, the Open Society Institute, and the United Nations Development Programme. It was founded "for the creation and transfer of knowledge concerning e-governance, e-democracy, and the development of civil society."[106]

In other words, within less than a decade of Estonia's e-government platform running smoothly, its leaders are training their counterparts in countries like Kyrgyzstan, Macedonia, and Senegal. This evolution is likely the fastest example of a public entity implementing an entirely new type of online government system, and then immediately turning around to export it to other countries. More recently, Skype evolved from a small, obscure team of programmers who created a telephone application to a juggernaut that threatens the existence of the incumbent telecommunications industry.

In essence, the Internet has been able to flourish in Estonia because the nation's independence coincided with the arrival of the disruptive technology of the Internet. As Estonia had a strong educational and technical background that existed before the Soviet occupation and improved during the Cold War, it was easy for free Estonians to create their own technological destiny quickly. The fact that the society and the government are so small means that visionaries like Linnar Viik and Veljo Haamer have a disproportionate influence over their countrymen, and that their ideas can be implemented without any bureaucratic legacy to stand in their way.

BECAUSE HAAMER WAS the first person to implement a public WiFi hotspot in Estonia and make a point of sharing that access through the use of an obvious marker—the stickers and the signs—he often is the expert whom small businesses and others turn to when they want to implement wireless access in their location. Further, Haamer had the insight early on that WiFi would continue to get cheaper. He recognized early on that WiFi would also be a relatively small marginal cost on top of a café's existing Internet access. As a result, there was essentially no reason for cafés to charge for this service. Haamer argues that free WiFi should be treated as a common good that every café patron can share, like a daily newspaper, even if the owner must incur the cost.

This attitude of communalism and benefit is very different from the experience of WiFi in the United States, where even in the early days large entities like hotels, airports, and coffee chains like Starbucks instituted paid WiFi service. Today, most Americans are accustomed to paying exorbitant amounts for WiFi access—$6 for thirty minutes is not unheard of at an airport.

Because of Haamer, free WiFi is the rule rather than the exception in Estonia. In most of the rest of the world, largely the opposite is true. Today, there is free WiFi in just about every café in Tallinn, in the port of Tallinn, on the ferry to Helsinki, in the bus station, at the airport, on the commuter trains. There is hardly another city on earth where WiFi is so commonplace, and so free. This basic infrastructure is what makes applications like Skype and e-voting possible, as everyone can get on the Internet, from the remote forests to the city centers.

In addition to evangelizing, Haamer also acts as free community tech support. His phone number is plastered on many signs in the WiFi parks around Tallinn and in other places around the country. He often gets calls from elderly Russian women who have heard about this "thing called WiFi," but don't really understand it. Haamer gladly will go over to their apartment and introduce them to the concept. He doesn't charge for this service, preferring to take his payment in smiles and baskets of food. Sometimes he will meet Estonians in public cafés in order to fix their busted WiFi cards. He does so cheerfully and eagerly, anxious again to simply spread his message. The fact that one more person gets online using WiFi is enough for him. He only makes his money when doing large-scale installations on things like ferries or busses, where a company has the funds to pay for the few hours' work.

Being from a small country, it's only natural that one wants to have solidarity with one's countrymen and bridge the existing digital divide that exists between elderly Estonians and younger Estonians. Even Haamer's own father, an eighty-year-old man who lives alone in a farmhouse in rural eastern Estonia near the Russian border, still does not use the Internet at all. But Haamer has helped his father become accustomed to using a mobile phone, a small step away from using a computer.

Haamer is constantly thinking of new public places to bring Internet access to. Shortly after he got going on WiFi cafés and parks, he immediately turned his attention to three locations that haven't gotten much attention anywhere else: supermarkets, shopping centers, and hospitals. There's no obvious reason why one would want wireless Internet in a supermarket, except maybe for the store's staff to have an easier way to monitor inventory—and if you're going to go that far, you might as well let the public use it for free as well, says Haamer. There's also the idea of creating a "captive portal," where the supermarket might be able to dynamically create advertising for specials going on in the store on the first web page that a user sees. The potential in a shopping center is a little bit more obvious, as there are people sitting down to eat lunch or stop for an afternoon coffee. But hospitals aren't usually places where even in America one typically expects free WiFi access. In Estonia, Haamer has made this possible.

Lately, Haamer's attention has been on forms of public transit. He worked to get free WiFi on the electric commuter trains that go from the center of Tallinn to the suburbs, a project that was completed in late 2006. The following year, he worked with an Estonian bus company, Hansa Buss, to bring WiFi on its "business class" bus service to Riga, the capital of neighboring Latvia. In 2008, he successfully completed WiFi access on the train from Tallinn to St. Petersburg. He is also continuing to lobby the government to require that new buildings in Estonia have subsidized, low-speed DSL (128 kbps) as a free service for all citizens, so that even people without WiFi gear can have a base level of Internet access. Given that in 2009 neighboring Finland required that all residents have a minimum speed of 1 Mbps commercially available to them, which will be increased to 100 Mbps by 2015, Haamer now has new ammunition to prod his government. In the end, as someone who grew up under Soviet occupation, Haamer is well aware of Estonia's great need in recent decades for advancement and prosperity. He remains grateful that he was able to play some small role in its modern development, especially now that Tallinn has been named as a European Capital of Culture for 2011.

"I'm just happy that I have worked with the whole country," Haamer says. "I saw so many people in the U.S. that are creating similar things, and they don't have such a lucky chance like I have. They don't have a whole country. I'm happy to be useful here."[107]

TIMELINE

February 24, 1918: Estonia declares independence from Russia

1940–1941: Estonia occupied by Soviets

1941–1944: Estonia occupied by Nazi Germany

1944–1991: Estonia forced into Soviet Union

October 14, 1967: Veljo Haamer born in Kohtla-Järve, Estonia

August 23, 1989: Baltic chain takes place across Estonia, Latvia, Lithuania

March 3, 1990: Estonians vote for independence

August 20, 1991: Independence restored to Estonia

August 23, 1991: Veljo Haamer and Peeter Parvelo found Korel AS

October 21, 1992: Mart Laar begins first term as prime minister; lasts nearly two years

February 21, 1997: First public Internet access point opens in Tallinn

March 25, 1999: Mart Laar begins second term as prime minister; lasts nearly three years

March 2002: Veljo Haamer founds WiFi.ee

August 29, 2003: First version of Skype released

March 2007: Estonia holds presidential election online; first such election worldwide

April 27, 2007: Two-week cyberattack against Estonia begins

November 2006: President George W. Bush receives a Skype phone from Estonian prime minister Andrus Ansip during a state visit to Estonia

May 2008: NATO Center for Excellence in Cyberdefense opens in Tallinn

4

Iran

Had nations better understood the potential of the Internet, I suspect they might well have strangled it in its cradle. Emergent technology is, by its very nature, out of control, and leads to unpredictable outcomes.

−William Gibson

The Arrest

Tehran

October 10, 2004

Omid Memarian finished a quick chat with a coworker and turned around to walk the few steps back toward his desk in the office of the Iranian nongovernmental organization Volunteer Actors. Looking up, about to enter his own office, he saw two men waiting for him. They had close-cropped beards and wore suits, Iranian-style, without ties. They spoke sharply and directly, without so much as a *salaam*.

"Please come with us downstairs to respond to some questions," said the leader in a flat, stern voice.

"Why should I come with you?" Memarian responded indignantly.

"You know about all the shit that you have done," the man said, biting into each word with a staccato tone. "Don't play games with us."

"Do you have a judicial order—a warrant?" Memarian cried out, his voice strengthening, with a tinge of fear. "Who are you? Why should I come with you?"

Memarian was also a blogger, and one who had used the new medium to publish a number of potentially incendiary posts that no Iranian newspaper would dare touch. Many of the posts covered city politics and local corruption—by Western norms, these writings would have been standard fare for any local media outlet. But in Iran, the red lines were constantly shifting, particularly with more and more liberal journalists and writers turning to the web as traditional media got shut down.

Memarian is among the first generation of Iranian bloggers, an entire generation emblematic of twenty-first-century Iran who are young, educated, and wired. Nearly 70 percent of the population is under the age of thirty, the average person has completed high school (many also go on to universities, where women outnumber men), and roughly 35 to 40 percent of the population is online.[1] As a result, this Internet-savvy, socially repressed, economically depressed, politically active generation grew up as the infrastructure and tools of the Iranian Internet matured, and it was only a matter of time before the two joined. This union created an Internet whose most important application was fighting back against the establishment, which in turn forced the Islamic Republic itself to use co-optive and coercive means to try to seize control of its online community. Ultimately, in extreme times, it has resorted to authoritarian means, including interrogations and jailings, to keep online dissidents at bay.

Memarian also knew that eighteen months earlier his good friend and fellow journalist Sina Motalebi had become the first blogger anywhere in the world to be arrested for what he wrote. Memarian spoke out against Motalebi's arrest on his blog.

Although Memarian didn't know it at the time, in the weeks before the bearded men arrived at his office, many of his journalist and blogger colleagues had been arrested under similar dubious circumstances in other parts of the city. In fact, a friend had called Memarian two days earlier to find out if he had been arrested. Memarian suspected that there was a possibility that he himself might be taken into custody, too.

In addition to being a journalist, Memarian ran Volunteer Actors, a nonprofit that provided training to other grassroots and civil society nonprofits in Iran. While Memarian stared down these mysterious men, his coworkers leaned in at a distance, trying not to attract their attention. The men's presence was enough to stop any movement in the office. Faces went silent as confusion and fear sank into the other five employees.

The second man slid the lower half of his suit jacket along his waistband, revealing a cold, metal pistol holstered at his side, and glared at Memarian. The first man stood by silently and waited for Memarian to comply. That was his answer.

"Bring your bag," the first man said, gesturing to the shoulder bag resting up against the foot of Memarian's desk.

He thought fast. There wasn't any incriminating information on that laptop, but it did have some e-mails, photos, and other documents—including a recent invitation to the United States and the Netherlands—that these men might be able to twist out of him.

"It's the office's bag. It's not mine," Memarian said quickly.

The ploy worked, and moments later Memarian followed them quietly out of the office, trying not to show how frightened he was becoming. He felt as if he needed to defuse the tension among his coworkers. After all, he'd helped to start the organization and didn't want any of them to get dragged into this situation.

The first man led the way down the three flights of stairs, and the second man followed behind. He could feel the second man's eyes piercing the back of his neck. He had never been held at gunpoint before. As they marched him out of the office, the leader quickly produced a list of names, with Memarian's on it, and barked more orders. When the trio got to the ground floor, the two men paraded him through the lobby and onto the street. Two more bearded suits were waiting, flanking both sides of the main door, and as the three passed they fell into position behind the group.

As the quintet approached a parked and waiting black car and van, Memarian could see that there were two more suits to the left, a few hundred feet away, standing in the middle of the street, and another pair to the right effectively barricading the other end of the block. Someone had sent a total of ten men, at least one of whom was armed, to apprehend one thin, thirty-year-old, clean-shaven journalist. Memarian was an active member of civil society, a thinker, a budding politician, a journalist, and, perhaps most dangerously, a blogger. He didn't think he could outrun any of them.

After seeing this, Memarian turned to the first man, who seemed to be in charge, and said incredulously, "Do you really think I'm going to escape?"

The leader ignored the question and spoke to the driver of the car. "We got him," he said. "You can go."

Then the leader whipped out a walkie-talkie and spoke quickly to someone, presumably at the men's headquarters: "We arrested him. Omid Memarian."

Only after this transmission did Memarian realize how serious things were getting. He had no chance to mentally prepare for whatever was about to befall him, or to alert his family and his friends. Put simply, they—whoever "they" were—wanted Memarian to disappear.

THE GROUP LED HIM into a large van with three rows behind the driver's seat. The first two men sat up front, with the driver, while one of the doorway guards sat in the second row, and another in the last row. They put Memarian in the third row, where all he could see were these men and tinted glass. No passersby could see that he was inside.

"What am I supposed to tell you?" Memarian asked flatly. "What is the problem?"

As the van began to pull away from the curb, the leader turned back to look at him.

"You have the answers," he said. "You know the problem. You will tell us everything."

As he spoke, he motioned for Memarian to hand over his mobile phone. Memarian did.

The first stop was Memarian's parents' apartment, nearly forty-five minutes away from downtown Tehran, in Lavizaan, a middle-class neighborhood.

Like most grown but unmarried Iranian children, Memarian still lived with them. Although his parents were not professionals, they had encouraged him to study engineering at university, where Memarian eventually earned a degree in metallurgical engineering. However, as he grew older, he spent more of his free time writing poetry and composing articles for reformist newspapers like *Hayat-e Noh*. His parents supported his transition.

By the time they pulled up to the apartment block, Memarian didn't know how he was going to tell his family what had happened. His first thought was that the armed man was going to frighten his elderly and ill mother, Touran. Memarian requested that they leave the gun-toting man in the van while the three others went upstairs with him to the fourth-floor apartment. The leader told him that they all had to go, but the armed man agreed to not enter the apartment.

As Memarian rang the doorbell, his mother answered the door dressed in *hejab* and carrying her *tasbih*, her Muslim prayer beads. When the door opened, her eyes rose to meet her son's face. Before they could exchange words, she saw the four men behind him. She knew that something was deeply wrong.

"Mother, these men are going to come inside to confiscate my things," Memarian said, trying desperately to be reassuring.

"Why is this happening? What have you done?" she cried out. But before Memarian or any of the other men had a chance to get a word in, she turned her growing rage toward the suits.

"My son has done nothing!" she said, looking straight at the leader. "Why do you do these things to our youth? What you're doing is crazy! My son has done nothing!"

"*Khanoom,*" the leader said in a quieter voice, trying to calm her down. "Please understand, we're just doing our jobs. *Inshallah,* he's done nothing wrong." This was the only instance when he showed any inkling of compassion toward Memarian or his family.

"Mom, it's going to be okay," Memarian added. "I've done nothing wrong."

"It's not okay!" she yelled out, her voice beginning to crack. "You know it's not okay!"

"It's a misunderstanding," Memarian tried to say. "I'm going to go with them, and will come back soon. I'm so confident in what I've done, nothing is wrong. It's definitely a misunderstanding."

But she wouldn't believe a word of it. Memarian could see it in her eyes that no matter what he said, she was going to assume the worst.

While the leader brushed past her and Memarian led him to his bedroom, his mother calmed down a little bit, and whispered prayers to God to not forgive these men for their crimes and to help her in getting rid of them. The other two men waited in the hallway.

"When will we be rid of these people?" she asked God, weeping and collapsing into a lounge chair. "They've made our lives difficult!"

Her exclamations then turned to the men.

"How much do you have to hurt and abuse our children?" she said, her voice faltering into sobs, "Why are you doing this to us? Why are you doing this to our children? These kids haven't done anything." She repeated these questions to herself over and over again.

Memarian had never seen his mother in a state like this, where she was unable to help him and was so obviously defenseless. As Memarian watched the leader flip through books on his wall-to-ceiling shelves, he could hear his mother's quiet prayers between gasps and sobs. The leader found articles that Memarian had printed from the Internet, copies of his old newspaper articles, photographs, an address book. He stopped to read some of it, tossed aside other papers, and stuffed what he wanted into a trash bag. Memarian remained silent as the man picked up anything that could potentially incriminate him. The leader, fastidious and deliberate, continued to read silently, ignoring Memarian's occasional pleas.

The leader was done after two hours, and he clutched the bag like prized treasure and went back into the main room, where Memarian's mother was still sitting. The leader tried again to reassure her, but she just sat, nearly expressionless, save the dried tears on her face. Memarian followed the leader out of the apartment, where they retrieved the other men and got back into the van.

The group sped toward downtown Tehran for what seemed to be forty-five minutes before stopping at a police station. Before Memarian was let out of the van, a police officer came to meet them and spoke with the group's leader. Apparently they'd forgotten to get Memarian's computer. All the men got out of the van, but instructed Memarian to stay put. Before he could react, two new men, including a driver, got into the van. The man with the gun returned and sat next to Memarian. The van turned around and they went back to the apartment.

When Memarian found himself with these three men at the door to the apartment nearly an hour later, he asked that he have two minutes alone with his mother, as she was emotionally vulnerable and ill with cancer. She was due to have an operation in the coming weeks, he explained. He would go inside, get the computer, say a few words to his mother, and would come out again. They agreed.

Fortunately, Memarian's sister-in-law happened to be there. When he saw her, he took a chance, grabbed a piece of paper, and rushed to write down the passwords to his various online e-mail accounts.

"Delete all my e-mails," he instructed. Memarian knew that he was going to be taken to prison, and one of the first things that his interrogators would demand would be the passwords to his e-mail accounts. While he didn't know exactly what would happen to him, or where he would be taken, he knew that his friend Sina Motalebi had returned from prison far more paranoid and far less gregarious than he had been before going in. Deleting whatever they might consider as "evidence" could possibly reduce the harsh treatment that would surely await him. He didn't have to say anything else, as she knew what had happened. She took the paper and nodded. He went to the bedroom, unplugged the computer, cradled it in his arms, and carried it out without saying anything else to his mother.

He rejoined the three men, they loaded the computer into the van, and then drove back into the city. At first Memarian thought that they were going back to the police station, but that thought was quickly chased away when the man with a gun took a blindfold out of his pocket. He instructed Memarian to put it on, lie down on the floor in front of him, and then cover his head with a blanket. As soon as he did, the armed man put his foot, in a thick black leather shoe, up against Memarian's throat.

The van drove around the city, circling a few blocks, sometimes stopping to go in reverse, and then resuming its course. The city's pollution, the dirt of the van's floor, and the sudden jerky movements made Memarian's stomach turn, but he knew that the best thing to do was to stay quiet and motionless. All of this driving was meant to disorient him. Memarian figured that they were somewhere in the middle of Tehran, simply based on a rough estimate of the time that they had been driving from the outer periphery, where his parents lived. But in a city of 12 million, that could be anywhere.[2]

"Cause of Iranian Government Policy, access to this site is forbidden."

Memarian's blog is but one of a community of many thousands. Currently, it is unclear precisely how large the Iranian "Weblogistan" actually is. Various estimates have put the number from 60,000 to 100,000, out of approximately tens of millions of blogs worldwide. While even 60,000 blogs may seem small compared to the millions of English-language blogs, Persian only has around 90 million native speakers globally, compared with languages that have far more speakers, like Spanish, French, or even Hindi and Mandarin Chinese. However, there are almost as many Persian blogs as there are French and German blogs, languages that have two to three times as many native speakers.[3]

Most Iranians tend to use one of the many blogging platforms that support the Persian language. One of the most common sites is Google's blogger.com, but there are others, such as persianblog.com, blogfa.com, blogsky.com, mihanblog.com, and parsiblog.com. Whatever the figure, it is clear that a sizeable minority of blogs in Iran is the most recent and poignant example of the younger wired generation pushing back against the Islamic Republic. Iran has been connected to the Internet since 1992, but only since 2000 has there has been a nearly constant struggle between reformist intellectuals' online speech and the government's effort to clamp down on any content that it deems to be offensive.

THE LATEST EXAMPLE OF THIS CONFLICT occurred during the June 2009 presidential election. Prior to the election itself, as has been the case in the past, the regime allowed many foreign journalists—including one from Comedy Central's *The Daily Show*—into the country. However, in the immediate aftermath of the election on June 12, 2009, when the government reported that incumbent president Mahmoud Ahmadinejad had won reelection with 62 percent of the vote, the Internet exploded with accusations of widespread fraud. The online rage erupted into massive street protests, the likes of which hadn't been seen in Iran since the Islamic Revolution. Nearly all at once, the government kicked out foreign journalists, significantly reduced mobile phone service (twice shutting off text messaging entirely), and slowed the national Internet capacity from five gigabits per second to two gigabits per second—less than half of its normal speed.

As such, the best way to get information in or out of the country was through the (albeit slowed-down) Internet, through blogs and Persian-language sites like Balatarin.com, and through social networking sites like Twitter, Facebook, and FriendFeed. Prior to the election, the different campaigns had set up online presences as a way to amplify their message. In fact, there was such concern for various online tools that the U.S. State Department famously called upon Twitter on June 16, 2009—just a few days after the election—to move its scheduled maintenance time to when it would be night in Iran. Within a week of the election, Google and Facebook rushed out new Persian translations of their services.

However, the simmering tension boiled over on June 20, 2009. In the warm Tehran evening, a young Iranian woman, Neda Agha-Soltan, was on her way to a reformist protest when she got out of her car on Kargar Avenue. She was watching the surrounding demonstrators and was suddenly shot in the chest and killed, apparently by one of the nearby Basij militiamen. Her shooting and subsequent death was captured on two different amateur videos, which quickly spread virally around the Internet. In death, she became a icon of dissent and a rallying cry against the regime. Her passing even spurred many efforts internationally—including one called NedaNet—to help Iranians have more unfiltered access to the Internet.

The reformists, particularly supporters of Mir-Hossein Mousavi, used their own websites (like ghalamnews.ir), blogs, and social networks as a way to get around the government's stranglehold on traditional broadcast media. But Iranian officials have also begun to co-opt the Internet for their own use, by sending out confusing Twitter messages with conflicting information concerning rallies.

While many of Iran's most well-known blogs and bloggers, including Omid Memarian, tend to be secular, and to varying degrees politically reformist, they may not necessarily constitute a majority of the Iranian blogosphere. The Persian language blog community has a very conservative and strict element as well, some members of which are funded by the Islamic Republic. In fact, in December 2008 the Islamic Revolutionary Guard Corps announced that they would be launching an army of 10,000 bloggers staffed by the Basij paramilitary forces. Even President Mahmoud Ahmadinejad has a blog. Indeed, for every blog that speaks out in favor of democracy, and criticizes some aspect of the Islamic regime, no matter how small, there are now also blogs where the supreme leader is lauded, and the president praised. Just as the American blogosphere is diverse in covering mundane personal details, so, too, is the Iranian blogosphere, which covers topics as diverse as sports and Persian poetry.[4]

Only a very small minority of Iranian bloggers, reformist or otherwise, have encountered the harshness of the Iranian regime to the same degree that Memarian has. Indeed, his case represents but one blogger of a generation who has agitated the current government. For Iranian reformist bloggers like Memarian, writing online has become an indispensable tool for citizens who want to speak out against their theocratic government. Despite the fact that Seoul's tech neighborhood has a Tehran Street, many educated and upper-class Iranians simply do not have the luxury of spending their time playing professional computer games. In a society demographically dominated by an increasingly young, smart, and wired generation, Iran has become a laboratory for how a twenty-first-century society deals with a government that leads based on seventh-century teachings.

WHILE IT IS TRUE that the government has silenced many of Iran's most vocal Internet users, the fact remains that online access is ubiquitous and inexpensive. Around one-third of Iran's 65 million people are online, more than double the percentage of its largest neighbors, Turkey and Saudi Arabia.[5] This figure is also bolstered by the fact that Iran's population tends to be fairly well educated, with a literacy rate of 77 percent. Indeed, an illiterate population cannot use the Internet.

Iran is furthering its efforts of technological progression since it first connected to the Internet in 1992. Most recently, in 2005, under the Khatami administration, the government completed an eleven-year, $700 million project on the construction of a 56,000-kilometer nation-wide network of very high-speed fiber-optic cables. "Enjoying the network, Iran now is ready to be the regional communication hub," President Mohammad Khatami said at the network's inauguration in July 2005, according to an account by the Associated Press.[6] "From now on we are capable to provide telecommunication services to the world."

While Iran may be an authoritarian nation, its behavior of investing in Internet infrastructure and allowing the price of Internet access to fall to the point where most people—particularly the burgeoning younger generation—can now have access whenever they need it indicates that it believes that this level of connectivity is necessary for the country's advancement. Of course, there is also the possibility we are simply witnessing unintended consequences. Initially, in the 1990s and early 2000s, the Islamic Republic probably had no idea that the Internet could eventually be used as a tool to oppose them in any meaningful way. Now that more and more people are gaining Internet access at school, at home, or at their workplace, there is that much more of an opportunity for citizens to encounter media that the Islamic Republic doesn't like.

THIS ATTITUDE IS IN STARK contrast to the way other authoritarian nations, including Burma (Myanmar), China, and Cuba, approach the Internet. The Caribbean island nation is notorious for keeping the price of Internet access so high and so restrictive that only 2 percent of the country is online. According to Reporters sans Frontières (RSF), ordinary Cubans who want to get online must wait for the better part of an hour to go to a *Correo de Cuba* (Cuba Post) bureau that provides Internet access.[7] There is a two-tiered pricing system, with international access costing $4.50 per hour (half the average monthly wage) and the national network, which most Cubans opt for, that costs $1.50 per hour. The latter service only allows for the sending of e-mails, and not web browsing.

Access to the international Internet from a public location, such as a Correo de Cuba, is highly monitored. While the connection is not censored as such, some type of software, likely a keylogger, monitors everything being typed. RSF adds that once a banned word is detected, an alert system pops up, and closes that application immediately.

In Myanmar, a few hundred people, primarily regime officials and military officers, strictly control the Internet. They determine who is allowed to go online, and who is not—so out of a population of nearly 50 million, only about 25,000 people are actually online. The Burmese government has created a strict policy that essentially keeps the Internet out of the hands of significant portions of the country. Further, the military junta has banned online political material and any "online material considered by the regime to be harmful to the country's interests and any message that directly or indirectly jeopardizes government policies or state security secrets." During Burma's massive protests of late 2007, the state famously cut off Internet access entirely for two weeks.[8]

IRAN'S HARSH INTERNET POLICY is probably most similar to that of China, although perhaps not as large, nor quite as sophisticated. Due to China's size, even though only roughly 25 percent of the country is online, there are over 300 million Internet users. (That figure far exceeds the entire global Internet population of 1997.) China also has a rapidly growing middle class and a corresponding rise in the number of university students. Indeed, by 2010, "Chinese officials estimate, at least 20 percent of high school graduates will be enrolled in some form of higher education; that number is expected to rise to 50 percent by 2050. China currently [in 2005] has about 20 million students pursuing higher education." As such, it is likely that in the coming years even more Chinese will be logging on to the Internet. In April 2008, China surpassed the United States in having the largest group of Internet users from any single country.[9]

The Chinese government has created the most draconian policy against the Internet in the world. Informally called the "Great Firewall of China," it consists of a series of routers, gateways, and other network devices that filter out unwanted material, while a team of censors makes sure that the lists are as up-to-date as possible. But the system is set up to be far more insidious.

As Clive Thompson wrote in the *New York Times Magazine* in 2006:

> The Chinese system relies on a classic psychological truth: self-censorship is always far more comprehensive than formal censorship. By having each private company assume responsibility for its corner of the Internet, the government effectively outsources the otherwise unmanageable task of monitoring the billions of e-mail messages, news stories and chat postings that circulate every day in China. The government's preferred method seems to be to leave the companies guessing, then to call up occasionally with angry demands that a Web page be taken down in 24 hours. "It's the panopticon," says James Mulvenon, a China specialist who is the head of a Washington policy group called the Center for Intelligence Research and Analysis. "There's a randomness to their enforcement, and that creates a sense that they're looking at everything."[10]

The OpenNet Initiative (ONI), a cooperative project between various universities in the United States, Canada, and the United Kingdom, studied Internet filtering in China from 2004 to 2005 and concluded: "China operates the most extensive, technologically sophisticated, and broad-reaching system of Internet filtering in the world. The implications of this distorted on-line information environment for China's users are profound, and disturbing."[11]

IN ANOTHER STUDY from the same year, ONI found strikingly similar conclusions concerning censorship in Iran:

> Iran has adopted one of the world's most substantial Internet censorship regimes. Iran, along with China, is among a small group of states with the most sophisticated state-mandated filtering systems in the world. Iran has adopted this extensive filtering regime at a time of extraordinary

growth in Internet usage among its citizens and a burst of growth in writing online in the Farsi language. . . .

Iran follows a pattern uncovered by previous ONI research, particularly in the case of China, whereby filtering regimes that reach a high level of sophistication target for censorship local language content and new forms of online expression, such as blogs. In instances such as Iran and China, the state demonstrates its commitment to a censorship regime that keeps up with the changes in technology. Such a filtering regime is effective in part by keeping citizens guessing as to how the blocking will work over time and introducing uncertainty into the equation. The OpenNet Initiative has found that a growing number of countries, including Iran and China, are shifting the Internet censorship regimes inwards using increasingly fine-grained methods of information control.

China, like Iran, has been quick to silence and expel its dissident writers. Yet it allows a proliferation of nonpolitical speech online. There is a tacit tolerance of blogs and of bloggers, so long as they do not stir up political muck, as some like Memarian have in Iran.

THE PROCESS OF ADAPTING MODERN communications technology is not something that is inherently new to the Internet era. Throughout history, Iran has acquired new types of communication technology, imposed restrictions on it, and then eventually adopted or perhaps even co-opted it. As Shahram Sharif, the head of a Persian-language technology news website called ITIran.com, put it:

> If you look at all the technology that's come into Iran, there's always been some sort of struggle toward it. When the telegraph entered Iran, it was against the law—you couldn't use it. Later, its use became okay—and then the fax, no one could fax anything. Then video cameras were against the law, and then they became available.[12]

In other words, the citizenry has always been able to stay one step ahead of government restrictions. Even Ayatollah Khomeini was able to smuggle in audiocassette recordings of his sermons while he was in exile during the 1960s. More recently, there have been on-again-off-again crackdowns on satellite dishes. Yet satellite dishes remain a fixture of the Tehran skyline. By all accounts, they are commonplace, despite the fact that they alone are responsible for the airing of television programming ranging from *Oprah* to CNN International to BBC Persian.[13]

Similarly, despite Iran's best efforts to stifle Internet access, determined users will always find a way online. Today, there is a large network of constantly changing lists of proxy servers that allow people to get around the filtered Internet in

Iran. A proxy server is an intermediary step that can be configured in a matter of seconds on a web browser by even the most nontechnical Internet user. There are sites that have long lists of proxy servers, which can fool sites into thinking the traffic is originating from somewhere other than inside Iran, and thereby will circumvent the filter. In fact, many public cybercafés, known locally as "coffee nets," will advertise (some with more subtlety than others) how to use "proxy servers" to subvert the authorities. This is illustrated by the fact that at many cybercafés in Tehran there is usually at least one person watching a video on YouTube, which until February 2009 was theoretically banned.[14] (YouTube was blocked again after June 2009.)

Another way that people can get around the state's web filtering is through extended use of RSS, an online syndication service that makes reading different websites more like reading them in an e-mail application. A web-based RSS reader like Google Reader will "ping" a site to check if there is new content, and if so it will automatically download that content. Google Reader, for example, fetches the content from Google's own servers, which cannot be blocked by Iranian authorities.

Further, there are efforts by various pro-democracy groups in the West to educate Iranians on how to start blogs and then maintain them by being cautious online. In 2005, Reporters sans Frontières published a "Handbook for Bloggers and Cyber-Dissidents" in multiple languages, including Persian.[15] This document provides explicit instructions for how to stay anonymous online, mainly by using a piece of software called Tor, which masks the user's point of origin. In other words, instead of making it seem that a blogger is coming from Tehran, if she's using Tor, it may seem that she is coming from another part of the world.

After the June 2009 presidential elections in Iran, Tor's executive director, Andrew Lewman, said that there were eight thousand users from Iran at any given time—a huge jump from the average of two hundred Iranian Tor users that his organization observed prior to the election. Another antifiltering software, Psiphon, reported in the week after the election that it noticed one new Iranian on its network every minute.

However, it is not clear how many average people in Iran are aware of these types of more complex instructions on filtering circumvention, nor if these techniques are having any sort of sustained effect on Iranians' unfettered access to the Internet.

IN DECEMBER 2000, dissident Grand Ayatollah Hossein-Ali Montazeri published a six hundred-page Persian-language memoir on his website, montazeri.com, while under house arrest. In this manifesto, Montazeri, a former architect of the Islamic Revolution, criticized the very foundations of the Islamic Republic that he helped to create, and accused Ayatollah Ruhollah Khomeini, the founder of the Islamic Republic, of executing thousands of political opponents in 1988. The government responded by setting up a similar site at montazery.com, which denounced Montazeri and represented the views of Supreme Leader Ayatollah Ali Khamenei. This first known overt crackdown on Iranian online speech came in December 2000. It represented a turning point, as this was the first time that the government directly responded to an online "threat" from someone else.[16]

By the next year, the Iranian government took far more serious action against online political speech in Iran. In early May 2001, the state shut down four hundred cybercafés in Tehran, under the pretext of their not having proper licenses. In June of the same year, the Iran Telecommunications Company banned children under the age of eighteen from accessing the Internet at all. Further, by November, the Supreme Council for Cultural Revolution declared that all Internet service providers should be controlled by the state and that ISPs were required to remove any sites deemed "anti-government" or "anti-Islamic." In September 2002, Judiciary Chief Ayatollah Seyed Mahmoud Hashemi Shahroudi called for the creation of a "special committee for legal investigation on Internet-related crimes and offenses."[17]

Today, the regime blocks hundreds—perhaps even thousands—of sites, including one of the world's most popular sites for video-sharing, YouTube. In Iran, if you try to access YouTube, you'll get a message in Persian and/or English that will inform you that access to the site is forbidden. This blocking is common for a whole host of other sites—pornographic, dating, political (both domestic and exiled)—as well as many blogs.[18]

OF COURSE, PULLING the country's plug on the Internet altogether would represent an unacceptable level of censorship, a fact not lost on the Iranian government. In late 2006, Iranian ISPs were told to throttle back the country's bandwidth to 128 kilobits per second to prevent people from using high-bandwidth online services, such as YouTube, Google Video, and Skype.[19] At that speed, it also makes it more difficult for Iranians to download movies and music, which may be identified as causing potential "harm" due to its Western origins. The measure was lifted in early 2008.

In January 2007, the Iranian government took yet another measure aimed at the country's blogger community, requiring Persian-language bloggers in Iran to register with the Ministry of Culture and Islamic Guidance, and provide their full name, address, e-mail address, phone number, and blog login and password. This move led to an outcry among the blogging community, with one blogger, Parastoo Dokouhaki, spearheading the "I will not register my site!" movement by creating online protest banners for other bloggers to copy. The government measure was due to take effect on March 1, 2007, but it remains unclear if any bloggers actually complied with it, or if the government has ever actually attempted to enforce it.[20]

The Internet Comes to Iran (1989–1997)

While the technological cat-and-mouse game between the Islamic regime and its opponents has intensified over the last decade, Iran has been online for nearly twenty years, the second-longest of any country in the region. Because the Internet has had time to mature in Iran, the country has one of the highest rates of Internet penetration in the Middle East, at an estimated 35 percent. Today, Iran has the fastest growth rate of Internet users of any Middle Eastern country, growing from 1 million Internet users in 2005 to 23 million in 2008.[21]

But this high level of connectivity did not happen all of a sudden. The complex online conflict in Iran is a direct result of the maturity of Iran's Internet infrastructure and history. Following the Islamic Revolution of 1978 and 1979 and the U.S. embassy hostage crisis, Iran was essentially barred from participating in any kind of American government research, including that of the ARPANET and its related projects. Further, by 1980, Iran's border dispute with Iraq had escalated into a very bloody war with its neighbor. The combination of these two events made the importation of any sort of technology transfer or training nearly impossible—Iranian educational and technological prowess was entirely devoted to the conflict.

With the end of the war came the first opportunities for Iranians to study abroad, and learn about advances that had taken place over the last several years, including the Internet. One of the first Iranians to use e-mail was Dr. Siavash Shahshahani, a forty-six-year-old mathematician who was a visiting scientist at the International Center for Theoretical Physics (ICTP) in Trieste, Italy, during the academic year 1988–1989. At that point, e-mail was one of the main applications of networking on the ARPANET, and had rapidly spread through the academic community. Shahshahani observed that e-mail was mainly being sent from scientist to scientist to organize professional work, to conduct correspondence concerning lab results, and to perform general day-to-day operations. Although skeptical of this new "gadget" at first, he soon became a "very active user" of e-mail. However, before leaving for Italy, Shahshahani had been appointed as the deputy director of the Institute for Studies in Theoretical Physics and Mathematics (IPM), a post he held for thirteen years.

In the spring of 1989 the newly appointed director of IPM, Mohammad-Javad Larijani, visited Shahshahani and other Iranian scholars and scientists at ICTP Trieste. Shahshahani and others lobbied Larijani to consult with the director of the Center, Abdus Salam, about how to get e-mail access in Iran. Abdus Salam introduced IPM to EARN (the European version of the academic BITNET), and by the fall of 1992 Iran took its first baby step on EARN. Over the next year or so, as the technical staff at IPM learned more and more about the nascent Internet, Shahshahani helped them to establish the first Internet connection to Iran.[22]

It is somewhat ironic that a technology that has caused so many headaches for the Islamic Republic can be traced back to a member of one of the most politically powerful and religiously conservative families in Iran today. Mohammad-Javad Larijani's brother, Ali Larijani, was the Iranian chief nuclear negotiator and is currently Speaker of the Iranian Parliament, and his other brother, Sadeq Larijani, is the new head of the Judiciary. Today, Mohammad-Javad Larijani himself is an advisor to Supreme Leader Ayatollah Khamenei and heads the human rights council in the Judiciary.

With Iran's newfound connection to the Internet, it wasn't long before commercial Internet providers began connecting ordinary people to the young World Wide Web. Unencumbered by traditional laws and a government that wasn't as aware of the changes afoot, Iranians began to swear on the public web, post "indecent" photos of themselves, and chat and flirt with members of the opposite sex. These, of course, were all activities that, without a computer screen to hide behind, Iranians would never publicly engage in.[23]

Three years later, in 1995, there were nearly 30,000 Iranians online, all of them getting their connection through IPM and its node access in Europe. At the time, Iran had more Internet users than any country in the Middle East except Israel. Having a single, painfully slow 9600 baud connection as its only link to the outside world made Iran very vulnerable to outages, as was shown for a couple of months in 1996, shortly after Congress passed the Iran and Libya Sanctions Act of 1996, which established economic sanctions on firms doing business with Iran and Libya. At the time, a "patriotic" National Science Foundation midlevel employee was single-handedly able to prevent Internet traffic from Iran—web and e-mail alike—from being routed into the United States for several weeks.[24]

AS MORE AND MORE IRANIANS were going online, creating their own virtual agora where previously forbidden issues could be discussed openly, Iranian politics was progressing along a parallel track. In 1997, Mohammad Khatami was elected as the fifth president of the Islamic Republic. He surprised many outside observers by winning as a moderate, despite his extensive conservative and religious credentials. After winning, he was hailed as a new leader who could balance and navigate Islamic theology and Western philosophy.

One of the most salient changes Khatami made to the Iranian political system was the expansion of civil society and social freedoms. In fact, he became the first Iranian president to ever use the phrase "civil society" in a national address, as he did during his inaugural speech before the Parliament on August 4, 1997:

Protecting the freedom of individuals and the rights of the nation, which constitute a fundamental obligation of the President upon taking the oath, is a necessity deriving from the dignity of man in the Divine religion. . . . [It requires] provision of the necessary conditions for the realization of the constitutional liberties, strengthening and expanding the institutions of civil society [jame'eh-ye madani] . . . and preventing any violation of personal integrity, rights and legal liberties. The growth of legality [qanun-mandi], and the strengthening and consolidation of a society based on a legal framework for conduct, interactions and rights, will provide a favorable framework for the realization of social needs and demands. . . . In a society well acquainted with its rights and ruled by law, the rights and limits of the citizens [shahrvandan] are recognized.[25]

In an interview with CNN, directed at the American people in early 1998, Khatami spoke extensively about a "dialogue of civilizations," and encouraged discussions with the United States in a conciliatory tone that has not been struck by any previous or subsequent Iranian president.[26]

Khatami didn't just talk about expanding civil society—he spearheaded various legislative efforts that made it easier for nongovernmental organizations to operate in Iran. While some limited relief and charity NGOs existed both before and after the Revolution, in the wake of Khatami's election, new NGOs that focused on a much broader range of issues burgeoned in Iran. For example, the number of NGOs with a focus on women grew from thirty to nearly six hundred, while environmentally focused NGOs went from about fifty to over five hundred.[27]

Khatami also sought to tone down the violent rhetoric that many outsiders are familiar with, including the "Death to America!" chant that many Americans have heard since the Islamic Revolution. At a Tehran rally two years into his presidency, in response to supporters cursing his opponents, he said, according to an account in the New York Times:

> "No, no, I don't like to hear slogans like that," he exclaimed. "I don't like to hear 'Death to opponents' or death to anybody, because as matters stand in our society at present, it will be interpreted in a very negative way, as meaning that anybody who does not share your views should be silenced, and that's not right at all. The Iran we want should be one where there will be room for all the different viewpoints, for all ideologies, even those that oppose the President. They, too, must have the right to express themselves."[28]

However, from the perspective of religious conservatives, Khatami's civil society agenda created a dangerous precedent where nonstate actors could freely engage in a civil dialogue with themselves and the government. Further, the restrictions on the press were much more relaxed than they had been before. This relative liberalization made the blogging that would come in the next few years finally possible.

The Rise of Reformist Media and Blogs (1998–2002)

According to Iran's 1979 constitution, media should serve the purpose of the Islamic Revolution, and "to this end, the media should be used as a forum for healthy encounter of different ideas, but they must strictly refrain from diffusion and propagation of destructive and anti-Islamic practices."[29] The government ensures that media is adequately Islamic by requiring that newspapers and magazines apply for publishing licenses from the Islamic Culture and Guidance Ministry. From 1979 until 1998, only a small number of licenses had been issued.

With the 1997 election of Khatami came a new minister of culture, Ata'ollah Mohajerani. The new minister accelerated the review process and issued more than five hundred new media licenses. Mohajerani also spoke out against censorship and allowed many translations of Western novels that previously had been banned. In February 1998, *Jome'eh* (Society) became the first reformist newspaper to publish, and within a few months there were fifty such papers. (Mohajerani now lives in the United Kingdom, where he writes a blog in support of the opposition Green Movement.)[30]

However, despite the reforms imposed by the Khatami administration, the presidency holds relatively little power in the Iranian political system. Ultimate authority lies with the supreme leader, Ayatollah Ali Khamenei, a conservative cleric. Khamenei's allies sit on many other unelected entities within the Iranian government, including the Press Court and the Judiciary. These institutions moved quickly to quash rising manifestations of civil society. *Jome'eh* was shut down by June 1998 after the Press Court claimed that it had published "anti-Islamic" articles. However, it reemerged as another paper called *Toos,* which was subsequently closed, and again it reopened under the name *Neshaat.* As such, this phenomenon came to be known in Persian as "chain" or "serial" newspapers.

One of the first victims of these censorship clashes was Faezah Hashemi, the daughter of former Iranian president Akbar Hashemi Rafsanjani. She was the publisher of a feminist magazine called *Zanan* (Women), and published an article accusing police intelligence chief Colonel Mohammad Naqdi of being involved in a 1998 assault on two government ministers. In December 1998, the Press Court found her not guilty of publishing false news, but did find her guilty of "insulting a senior police chief," and ordered her to pay a fine of $500. The magazine's operations were also suspended for two weeks in January and February 1999. Throughout the first half of 1999, many newspapers, including *Khordad* and *Hoveyat-e-Khish,* were also shut down and their editors imprisoned for "publishing lies" and other "anti-Islamic" offenses.[31]

In July 1999, another reformist paper, *Salam,* was ordered closed indefinitely. Its crime was having published a secret memo written by a former intelligence agent named Saeed Emami, who had provided his superiors with a plan on how to crack down on the reformist press. *Salam*'s editor, Abbas Abdi, was arrested, triggering massive protests in Tehran and throughout the country.

On September 5, 1999, *Neshaat* was ordered closed for "insulting the sacred decrees of Islam and the supreme leader," because the paper published an opinion piece that raised questions about capital punishment in Islam, while another challenged the authority of Khamenei. Its editor, Latif Safari, was forbidden from working as a journalist for five years and was sentenced to thirty months in prison.

It was in this political and journalistic environment that a twenty-year-old Omid Memarian began to form his own social and political thinking. He made a point to read *Jome'eh*, later *Toos*, and then *Neshaat* each day—and he began to imagine what he would say if he wrote for these newspapers; he was so taken with the reporting, particularly exposing political corruption and the publication of secret documents, that he brought copies of the newspapers to new NGOs that began forming in surrounding neighborhoods. Memarian began to gain a reputation as an organizer and someone who extensively participated in Iran's embryonic civil society.

As a high school student, reading newspapers was a bit out of character for Memarian, as he always excelled in math and physics. This is part of the reason why he ended up studying metallurgical engineering at Tehran University. Numbers just seemed to make much more sense than the nebulous meanderings of centuries-old Persian literature. He was so good in math, in fact, that as a university student he began tutoring small numbers of high school students.

One evening, a family friend was over at the Memarian apartment, and was speaking with Memarian's mother. He overheard the neighbor talking about how she wanted to hire Memarian as a tutor, but couldn't afford to pay his going rates, and was willing to barter her services as a cook or as a house cleaner in exchange for her son's education. Memarian was instantly overtaken by a sense of shame, and immediately wanted to tutor this student for whatever amount his family could afford.

The Memarian family lived in a huge apartment block that was mainly occupied by military officers and their spouses and children. Memarian figured that there must be plenty of high school students who, despite their fathers' decent military salaries, were unable to obtain an adequate education to score higher on the ever-competitive end-of-high-school exam known simply by its French name, *concours*, or competition. At the beginning of each new tutoring session, he would double-check with the students to make sure he wasn't causing undue financial hardship. Within a few months, Memarian opened a tutoring center at a local mosque.

Beginning with only a few teachers and several dozen students, the tutoring center quickly grew over the next few years to many times that number.

Memarian wanted to expand it so that these kids would have an opportunity that he didn't have, a place to hang out outside of the watchful eye of one's parents. The Center for Youth Alternative Thinkers, as it came to be known, became a sort of youth group with a loose religious affiliation. It provided a place for kids to go after-hours, where they could receive additional education at a low cost, and could also play games, hold discussions, and receive guest speakers. Such an organization is especially crucial in a country where half of the population is under twenty-five years old.[32]

Soon after, the center's council decided that it should begin publishing a newsletter to be distributed in the neighborhood, beyond the safe confines of the mosque community. The first issue had but four pages and contained articles about the goings-on in the neighborhood, cartoons, jokes, and poems. Memarian himself penned a lengthy piece praising the ideas and leadership of Ayatollah Khomeini, a piece that he wrote simply to placate the mosque's leaders, who wanted to make sure that the group stayed true to its religious roots.

As the months went on, Memarian began to realize that the Center for Youth Alternative Thinkers was outgrowing the mosque, and that with the concurrent proliferation of NGOs, it made more sense to split off and become an independent group. That way, they wouldn't be subject to content controls, and could even bring in speakers (such as those who would lecture on safe sex) that were reluctant to be in front of a potentially religious audience. The center, under the same name, was officially registered with the City of Tehran in late 2001.

When Memarian finished his university studies in May 2002, he took a job working for Clark Sanat, an Iranian company that manufactures gas, oil, and petrochemical filters, among other metallurgical products. But the job got to be rather tedious. It was hard to focus on working in a factory when his mind was on his kids, the newsletter, and who this week's guest speaker would be. While bored at work, he penned letters to the editors of many of the newspapers that he voraciously read. His letters mostly touched on political topics of the day.

Through the youth organization, he became part of the civil society that Khatami had spoken about. His youth organization served as a sort of forum to discuss various social issues of the day, such as job prospects and the role of religion in public life. Memarian's letters eventually turned into short op-ed pieces—the first of which was published in late 2001.

ON JANUARY 16, 2000, another reformist daily appeared on the streets of Tehran—called *Hayat-e Noh* (New Life). This paper was a bit different, as Hadi Khamenei, the half-brother of Supreme Leader Khamenei, was its publisher and main financial backer. The paper's editor-in-chief, Abbas Safaifar, told the Associated Press on the first day of publication: "Our audience will be mainly intellectuals and students. We emphasize freedom of thought and expression and religious tolerance."[33]

Managing editor Masoud Safiri wanted to combine the political weight of *Le Monde* with the populism of *USA Today*. As such, the paper took "soft news" topics and used them as metaphors to make political statements indirectly. The paper frequently made reference to Western authors and thinkers who spoke out against oppression, such as Nelson Mandela and Vladimir Nabokov. Once, the paper wrote an editorial calling for Iranian soccer star Ali Daei to "hang up his cleats" while at the top of his game in order to make way for the new and younger generation. Seemingly innocuous, this editorial was published around the time that former two-term Iranian president Akbar Hashemi Rafsanjani was seeking a new position as a minister of Parliament. Indeed, Iranians have learned both to write and read between the lines.[34]

Within a few months of the paper's opening, Memarian found himself being interviewed by the paper's social and lifestyle section. The paper knew about Memarian because of his work with youth and the burgeoning community of local NGOs. After being impressed by his dedication, the editor encouraged Memarian to submit opinion pieces about social issues, particularly ones having to do with youth issues and culture. He did so, submitting a handful of articles per week—and Memarian was thrilled each time he turned one in.

While *Hayat-e Noh* didn't pay for his contributions, Memarian took it as seriously as he did anything else that he had ever done. Now was his chance to get in with a newspaper, and work with the people whom he idolized, the people behind the bylines. They, in turn, were so impressed with his work that after a few months he was offered a part-time job with the paper. He didn't hesitate for a second. Despite the fact that the paper was only offering him a few hundred dollars a month—a fraction of what he was making as an engineer—it was truly where he wanted to be.

By mid-April 2000—just two months after an election in which the reformists won a decisive majority in Parliament—the conservative-dominated Judiciary came down hard on the community of reformist newspapers and magazines, closing fourteen papers in a single day. This was the second such major crackdown in as many years. All of the papers were charged with "continuing to publish articles against the bases of the luminous ordinances of Islam and the religious sanctities of the noble people of Iran and the pillars of the sacred regime of the Islamic Republic."[35]

Given the toxic atmosphere of publishing in Iran, one might think that the reformist journalists would have migrated to online publishing much sooner. After all, during these same years, many new online magazines, including *Salon* (1995), and *Slate* (1996) were founded in the United States. However, while the Internet was flourishing in the United States in the late 1990s, at the height of the dot-com boom, Internet access in Iran was still rare and expensive. Even though the Net came to Iran in 1993, six years later only 80,000 people (out of

65 million) were connected.[36] The country relied exclusively on external and expensive satellite connections via Canada, Europe, and other Persian Gulf states.

The few people who were connected to the Internet had very little material to read in their own language. This problem was complicated by the fact that there wasn't a standard way to display websites in the Persian language, which uses a non-Latin alphabet and is read right-to-left. Not all word processing programs and web browsers supported the various fonts that at best only half worked in Persian.

In 1991, a group of computer scientists got together to solve this problem of how to get different alphabets to display properly across different applications, different web pages, and different operating systems—they called it Unicode. Nearly a decade later Unicode became sophisticated enough that it included the Persian alphabet and was supported in both Microsoft Windows and its browser, Internet Explorer. (Today, Unicode is nearly standard in all browsers and operating systems, so it is easy to type and read in Persian, Russian, Chinese, or most other non-Latin alphabet languages.)

Once Unicode began to support Persian, Hossein Derakhshan, a young Iranian columnist who wrote about the Internet for *Hayat-e Noh,* took notice. On November 6, 2000, he wrote a column extolling the virtues of Unicode, and observed that this could bring about a potential radical change for written Persian online.

By the end of the year, Derakhshan had immigrated to Toronto with his Iranian-Canadian wife, where he immersed himself in online content on a much faster Internet connection than he'd had in Iran. He discovered some of the earliest generation of bloggers, including Jason Kottke and Dave Winer, who combined their interest in current affairs and technology in a way that Derakhshan wanted to emulate.

He continued writing for Iranian newspapers from his newly adopted home. As the months went by, he tried to use Unicode, but the computer he was using ran Windows 98, and the only program that supported Unicode was the stripped-down word processing program Notepad. Even when he could get an article typed and would send it to Iran, the piece didn't always come out exactly the way he had written it.

"After awhile, when I realized that writing from Canada and sending to Iran was such a difficult thing, they were censoring it, and they were mistyping the stuff I had written," Derakhshan said.[37] "It was horrible."

In the subsequent months, Derakhshan tried to publish some articles on different websites, but again, none of the sites made it quite as easy as he felt it should be. None of the sites allowed him to type directly in Persian, and display in Persian, as easily as could be done in English, or any other language written in the Latin alphabet.

As Derakhshan was exploring the nascent blogosphere in 2001 from Canada, so too was a twenty-one-year-old computer science student at Tehran's prestigious and technically oriented Sharif University. Salman Jariri spent much of his free time reading Kottke and Winer, just as Derakhshan was doing. He had also just ordered an e-book through the Internet, *The Lexus and the Olive Tree,* Thomas Friedman's seminal work on globalization.[38]

On September 7, 2001 Jariri started the first Persian-language blog using Unicode. He didn't make a point of drawing attention to himself, and didn't link to any other blogs. He alerted only a few of his closest friends and family members by e-mail to his new endeavor. As there was no existing blogging platform, he had to code each page by hand, a process that was tedious and time-consuming. His first piece:

> What is the meaning of a weblog? Weblog, website or homepage are all personal writings that are about an individual's interests and thoughts. Weblogs are updated every day. You can go to Google to see others' weblogs.
>
> The many interesting points that I see, read, or hear, throughout the day . . . or the interesting things I find on the web . . . to the thoughts and issues that come to my existence . . . everything![39]

Three weeks later, and completely unaware of Jariri's blog, Derakhshan himself started one of his own. He'd been living in Toronto for nearly a year, and immediately after September 11 the word "blog" entered the English-speaking lexicon, as some bloggers who had been writing in near-obscurity outside of a core community gained some recognition in the mainstream press. These included nearly the exact same set of blogs that Jariri had been reading—Jason Kottke, Dave Winer, and Jeff Jarvis.

"I started reading these and then I realized that this is exactly what I have to do now," Derakhshan recalled. "Because this gives me an amazing platform for my style of writing, for the stuff that I wanted to address and the content that I was actually already writing about in my columns when I was in Iran—they were so similar to blogs."[40]

He continued writing over the next few weeks, mostly commenting on the state of the nascent Persian blogging world. By early October, he discovered Jariri's blog and pointed out that Jariri claimed to be the first Persian-language blogger.[41] The following month, Derakhshan put together a definitive guide that outlined how to create a blog in Persian, using the free site blogger.com.

As blogging began to take root, editor Masoud Safiri encouraged his writers to follow Derakhshan's example as a blogger. He estimated later that within the first year, nearly 30 percent of the one hundred writers and editors at *Hayat-e Noh* had blogs.[42] Many journalists at other reformist papers, frustrated with the constantly fluctuating "red lines"—the ill-defined boundaries of censorship imposed by the Islamic Republic—turned to blogs as a way to skirt the rules. One blogger, Parastoo Dokouhaki, a twenty-six-year-old former journalist for the reformist and feminist weekly magazine *Zanan* (Women), once used her blog to describe a documentary film that she had seen at a conference that featured interviews with Iranian prostitutes—a subject that could not be mentioned even in a feminist magazine like *Zanan*.

Sanam Dolatshahi was one of Derakhshan's newspaper readers. From her family's apartment in Tehran, she used to cut out and save Derakhshan's writings about the Internet every week. She didn't always understand what he wrote—she didn't have a computer and had never sent an e-mail—but she had a visceral sense that they were important. The collection of articles would become relevant one day, she thought.

"I didn't know what the Internet was, but I knew that one day I would have it," she says, recalling how the tone of Derakhshan's articles gave a sense of urgency and importance.[43]

Sometime in the late summer of 2001, after Derakhshan had moved to Canada, Dolatshahi finally learned how to send e-mails. One of her first messages was sent to Derakhshan, and she asked where he was writing now that his columns had disappeared. He replied, saying that he had moved to Canada and that she should check out Gooya.com, a popular Persian-language news website run from outside Iran that had reprinted some of his columns. By November of 2001 she came across his online guide to creating a Persian-language blog.

Because Dolatshahi didn't know how to have multiple browser windows open at once, she printed the entire manual and put it in front of her keyboard and followed, step-by-step, his precise instructions.

Her first entry, written under the pseudonym *Khorshid Khanoom* (Lady Sun), was published on November 9, 2001:

Hi,

These are the first words of my blog. I should confess, since I don't know how to type in Farsi, I am killing myself to write this post.

Also, I don't like this font at all. I should ask Hoder how I can change the font.

But I will ask my question in the Yahoo group he has set up.

Well, you might want to know who I am. My name is Khorshid Khanoom. I'll turn 24 in a week. I don't know anything about computers. I'm an English teacher and a master's student in English Literature. I saw that women's place is empty among blogs and so I started mine. Of course I should say that I'm very busy, so I'll write once a week.

My blog's topics will be varied. I'll write about the things I like and I don't like. I hope you'll also help me and send me some content. Also, I'd like to know what women expect from a female blogger. For example, would they like to read about Brad Pitt, cooking, fashion, and stuff like that (well, maybe I'll write about Brad Pitt, but I doubt about the rest!) or, would they like to read about movies, music, literature, and stuff like that? Well, of course I'll do what I want to do, but I'm interested to know people's opinions.

This week's post was just a test. I'll practice typing in Farsi and will try to learn a bit about computers so that I officially start blogging next week.

Based on the unwritten Persian blogging principle, I thank dear Hoder for putting the blogging bug into our lives.

Also, I apologize from other bloggers for not putting the links of their blogs, because I still don't know how to do it! But I hope they will link to me (of course from next week.)

Please e-mail me at this address: [redacted][44]

By the springtime, she and a handful of other bloggers got together at a Tehran coffee shop to start a new online magazine called *Cappuccino*. Dolatshahi herself came up with the name after spotting an attractive man in the café who was ordering a cappuccino.

By August, Memarian followed *Cappuccino*'s example and began blogging about all of the things that *Hayat-e Noh* would not publish in print. He talked about political control, corruption, and influence peddling in local government. He never made attacks of any kind against the Islamic government or the supreme leader, which he knew would be pushing the limits too far.

Later that same year, in December 2002, one of Memarian's coworkers, a film and political reporter named Sina Motalebi, was also inspired by the efforts of *Cappuccino* and created a blog as well. He published his first article on December 1, 2002, about the trial of Hashem Aghajari, a university professor who criticized aspects of Iran's clerical rule and was sentenced to death the previous month.[45] This was one of the first instances in which a political reporter who had previously been restricted even by a reformist newspaper's censors and government "red lines" took to the Internet to publish politically sensitive material.

As Memarian said in a later interview:

I think I felt that, first of all, even for me, it wasn't clear for me what are the consequences of self-publishing. We were among the first people, [the first] generation. We were examining the self-publishing phenomenon. I couldn't understand that when I sit down behind my PC and publish something that it has consequences. We were just testing ourselves. I thought as long as I do not say—even on the blog—I never criticized the supreme leader. The major red lines I tried to avoid. [I thought that] as long as I do not lie and I do not spread inaccurate information, I can defend on what I say. I was careful not to say the things that I wasn't sure about. There are things, even though I told you that there are red lines that you know, there are red lines you don't know. It's always changing. You don't know if you're past the red line or not.[46]

Crackdown (2003)

Despite the fact that Khatami had been elected in 1997 and was reelected by a closer margin in 2001, Iranian conservatives continued to be a dominant political force. They control the unelected Council of Guardians, the upper house of the Iranian Parliament, which has the power to veto any bills that come from the lower and popular Majles. Further, the Council has the power to dismiss any candidates running for any kind of political office, local, regional, or national. Conservatives also retain control over the Judiciary and the Ministry of Intelligence, which have no oversight from the president or the popularly elected Parliament.

This tension between the reformists (who struggle for more social freedom) and the conservatives (who hold all the real authority and power in Iran) created a situation in which, despite Khatami's best efforts, the citizenry became more disillusioned about the lack of any substantial change. Indeed, from 1996 to 2002, Iran's unemployment rate nearly doubled, going from 9.1 percent to 16.2 percent.[47]

In early 2003 the conservative government began arresting individual journalists—and later, bloggers—with far more frequency than they had before. Ali Reza Eshraghi, the op-ed editor at *Hayat-e Noh,* was arrested after having authorized the publication of a 1937 American political cartoon showing an oversize thumb pressing down on a man in a black robe. The Islamic Republic somehow had determined that this cartoon, which originally depicted President Franklin Roosevelt's influence over the Supreme Court of the time, was in fact a veiled attempt to go after the founder of the Islamic Republic, Ayatollah Ruhollah Khomeini. The newspaper was immediately shut down, and Eshraghi was put in solitary confinement for two months and was only freed on $31,000 bail.[48] Memarian spoke out against the arrest of his coworker and friend and the shutting down of the newspaper on his own blog.

In 2003, local elections were held across Iran, and were seen as the bellwether for Khatami's chances at a second term as president. Things did not go well for the reformists—conservatives took fourteen of Tehran's fifteen seats for city council. Local voters had previously elected all reformists in 1999. However, with frustrations mounting against the reformists for not having made good on many of their promises of economic and social reform, many reformist-minded voters boycotted the elections entirely. Indeed, only 12 percent of eligible voters in Tehran showed up at the polls. Nationally, turnout was at 49 percent.[49]

Khatami himself called the high rates of abstention an "alarm bell for the future."

By early 2004, just prior to parliamentary elections, the conservative-led Council of Guardians had dismissed over two thousand candidates who had submitted their names to be on the ballot. This was four times as many names as the council had disqualified in the past, and included Mohammad Reza Khatami (the president's brother, and deputy speaker of the outgoing Parliament), who was among those who were disqualified for being indifferent to Islam and the constitution, or for questioning the supreme leader's powers.[50]

Starting in January 2003 and continuing through the weeks after *Nowruz,* the Persian New Year (March 20, 2003), Sina Motalebi was summoned to the prosecutor's office many times for questioning.[51] The authorities wanted to know about what he had written on his website about political dissidents. Memarian recalled that Motalebi would go for questioning nearly every day, and he would always call his wife immediately after he left. As these meetings became both longer and more frequent, he told his wife that if one day he didn't call her when he left the office then it meant that he had been arrested.[52]

On April 20, 2003, Motalebi didn't call.

Memarian immediately turned to his blog to express his contempt and demand Motalebi's immediate release. Not only was he outraged at the fact that Motalebi, his friend, was arrested, but he was astonished that the government couldn't come up with any reasonable explanation for the crackdown.

Memarian said in a later interview:

> I thought that in general, that we are the Children of the Revolution. You can accuse everyone of doing something against national security or trying to overthrow the regime. We are the Children of the Revolution—you can't do these kinds of things to us. We're paying the price for the low standards of living. We're the citizens of Iran.[53]

What was it that Motalebi and Memarian had written about on their blogs to trigger such a draconian response?[54] Motalebi's blog has been wiped clean from the Internet, so there is little record of his posts anymore. Some of Memarian's, on the other hand are still available on a blogspot.com page.

On January 12, 2003, the second day of Memarian's blog, he wrote:

> Newspapers can publish the statements of high officials without fear of being closed by the government. But we in *Hayat-e-No* Newspaper are afraid sometimes even of publishing even what President Khatami says. We sometime censored his words so as not to cause us any problem or prosecutions for ourselves. We were hardly trying not to write anything that would make an excuse for conservatives to sue us, even we used to modify certain words. For example, instead of "contradiction" we used to

write, "different viewpoint." It is so painful and brings tears to one's eyes being unable to write what you believe, but they [conservatives] even couldn't tolerate us and closed the newspaper.

A few months later, on April 25, 2003, he wrote just a few days after Motalebi was arrested:

Sina Motallebi went to prison too. . . . I remember it was early April 2003 that I found out in the middle of a busy day at the newspaper when [Ali Reza Eshraghi] came in and was silent. I asked him what had happened and he was hesitant to talk. I insisted. He told me that they interrogated him. Why? Didn't know. Well! Of course they asked him about everything but it was not clear what was his charge and this made him more worried. When they bring you to an interrogation desk and you do not know why, and they keep you hours and hours and make you walk with a disgusting laughter you will be afraid, you will have the detestable feeling of being like a submissive person.

Ali Reza Eshraghi was imprisoned before Sina. He spent more than 50 days in solitary confinement. In his last days he told his mother that he wished he was dead. I meet him two nights ago; he was not the same Ali Reza who had been working 20 hours a day, writing and mixing earth and heaven creatively together. He had totally shrunk and was very cautious. There was still generosity in his eyes but frequently was looking around and there were tears coming to my eyes. He told me that he wanted to quit writing for a while. "It's not worth it." I remember when he was writing and everybody was jealous about. When they arrested him, we were shocked.[55]

On May 5, 2003, he addressed some of Iran's main contemporary social problems:

The head of National Organization of Youth is speaking like Saeed Al-Sahaf [the former Iraqi Minister of Information]. . . . Everybody knows that youth tend to use drugs, are hooligans and riot. . . . Everybody knows that the average age of marriage has increased, everybody knows that AIDS is like a time bomb in our country, everybody knows that youth are tending to violence more and more. . . .[56]

In a later interview, Memarian said this of his posts:

We were talking about social restrictions, youth and their needs, and the society [where] 70 percent of the population is under 29. They need jobs, they want to be entertained, they have lots of social and political demands, and so the government cannot afford that. [We were] talking about youth and their needs, and criticizing the government's position against youth is something political. It has consequences.

Everywhere in the world, talking about civil society or NGOs or grass-roots activities, it's so normal; it's not sensitive at all. But in Iran, as the government is so sensitive about civil society ideas at different levels—they think that civil society is [an opening] for foreigners to come to the country and overthrow the Islamic Republic. When you talk about civil society organizations, they get sensitive. They think you are an agent, that you have a plan to spread civil society.

For us everything was very normal. For the case of [Motalebi], they were not very familiar with websites. It makes their efforts to control the information useless. They have no idea how the Internet makes their efforts useless. When they understood, then they knew the front line was going to be Internet.

They didn't know what they were doing. They were not prepared for that. That's why they arrested people one-by-one, they thought that by intimidating [us], the other people in this field would get scared and would stop their work. They will have time [to deal] with this new phenomenon.[57]

Motalebi was released from prison on May 12, 2003, and Memarian wrote about it on his blog. That night, Memarian and Ali Reza Eshraghi, who had just been released from prison a couple months before, went to visit him with a few other friends. They were led into the living room, where friends and family had gathered around Motalebi. Persian tea and sweets were passed around, but Motalebi didn't say much, nor did he eat. He sat back, almost trying to disappear. A normally gregarious and sociable person, it was clear that Motalebi was still afraid of what the government could do to him, even outside of prison. He didn't want to talk about his experience, nor did he want his friends and his family to get hurt in the future because of something that he said or did. When he spoke, his voice trailed off, and he stared into space, as if something on the opposite wall had suddenly stolen his attention.

Eshraghi, as someone who had also experienced Tehran's prison system, told him that the best way for him to heal was to stay busy, to continue working. Eshraghi pointed out that although *Hayat-e Noh* had been closed down, it still had a sister publication focusing on economics, and Motalebi would certainly be welcome there.[58] But Motalebi didn't seem to be paying attention, despite Eshraghi's insisting that Motalebi leave his family's house, get outside, and attempt to distract himself in some way, to accept what had happened and move on. Both Memarian and Eshraghi were worried about what might become of Motalebi and how he could continue to survive both professionally and personally.

"Sina was not that person that I knew," Memarian remembered later. "We knew that something breaks when you go there."[59]

The Second Wave of Arrests (2004)

On December 11, 2003, world leaders, policy wonks, and geeks gathered in Geneva, Switzerland, for the first World Summit on the Information Society (WSIS). This was the same conference where Senegalese president Abdoulaye Wade gave his speech calling for "digital solidarity." Geneva, of course, is no stranger to such gatherings—it hosts many of the world's international bodies, including many of the United Nations' auxiliary organizations. But this was the first UN-sponsored world conference to focus entirely on information technologies, namely, the Internet.

Since *Hayat-e Noh* had been shut down, Memarian had gone back to working for NGOs. He'd taken a full-time job with Volunteer Actors, an NGO that he'd helped to found, and aside from submitting an article here or there, he'd largely given up on writing. He attended the WSIS conference to network with other NGOs from around the world.

In the first day alone, the presidents of Finland, Azerbaijan, Egypt, Mali, and Latvia spoke to the assembled masses about the benefits of a connected world. President Khatami of Iran was among the scheduled speakers. Dressed in a charcoal cloak and black turban, he orated in an unanimated monotone to a largely sedate audience. His seven-and-a-half-minute address touched on the importance of the diffusion of this new Internet-based knowledge.

> The entry to the information society is a new opportunity for the entire world population. The "information age" is the "age of dialogue" and the "networked society" is the organizer of the "networked order." We must seek a solution and work out a formula so that "exchange of information" in the information society leads to "dialogue" and shortened distances. At the outset of this millennium, I raised the need for "dialogue among civilizations," in the age of cyberspace, too, we should continue to encourage and promote "dialogue among civilizations."
>
> The information society must take cultural diversity as the foundation for the common existence of human society and must be able to rely on it. We must work toward securing the participation of all cultural, social and linguistic groups in the creation of a knowledge-based society.
>
> The establishment and consolidation of knowledge-based societies requires commitment to ethical values, human rights and principles of democracy and instruments of "good governance." Politicians and experts should be able to delegate power to the people through the electronic process.
>
> We are concerned about inequalities in the development of infrastructures and global access to and use of information and communication technology. We should focus on the objective of turning the digital

gaps into digital opportunities through the promotion and consolidation of digital ties.

In this perspective, we shall strive and endeavor toward the fulfillment of rights such as the "right to development," the "right to communication" and the "right to information." We urgently appeal the international community to help create new capacities in the developing countries and assist them with their empowerment.[60]

Despite Khatami's appeal for a "right to communication," some BBC journalists attending the conference who had followed the rise of Iran's blogging community and were aware of Motalebi's arrest wanted to confront the president directly. The following day, Khatami held a press conference in an adjoining room before approximately thirty or forty reporters from Iranian and foreign media outlets. Among them were Memarian and Aaron Scullion of the BBC.[61]

Scullion and three other journalists from the BBC were part of a blog called "The Daily Summit," sponsored by the British Council. In the run-up to the WSIS conference, they had posted about the rise of Persian blogging, and drawn heavily from Hossein Derakhshan's blog for source material. The quartet of British reporters solicited questions from the Persian blogosphere to directly question Khatami—Derakhshan himself blogged: "This is the best evidence of how a simple blog, in the right time and right place, can be used as a strong tool for political change—this is [how] technology can help democracy. Isn't this what the whole summit is about?"[62]

For whatever reason, most of the questions directed at Khatami had little to do with the Internet or the "Information Society." After a handful of questions pertaining to Iran's relationship with Egypt and its role in the reconstruction effort in Bosnia, Scullion confronted the president head-on with this question: "Will you pledge uncensored access to the Internet or publish a list of sites deemed unacceptable?"

His response:

The BBC, Voice of America, and other American sites will not be censored in Iran. Many things that are contrary to the policies of Iran are available in Iran. Even opposition websites are available. We are exerting greater control over pornographic and immoral websites that are not compatible with Islam. And even some political sites that are very insulting to religion. But we are not censoring criticism. Criticism is ok. But all the sites and the numbers are very few are pornographic sites and even in the Western countries, they are, to some extent, controlled. Certainly, political sites insult in a very illogical way our religious beliefs, because this inflates the religious sentiments. We have our Minister, if you want him, he can come here and say how many are controlled and censored.

I don't think there are that many. He says that altogether there are 240 and the majority of them are pornographic sites, not political sites.[63]

For Memarian, and the other Iranian journalists in the room, Khatami's answer was typical—his meaning was shrouded in a duality that appeared to give a nod to both reformists and conservatives. By saying that "criticism was ok," and yet at the same time acknowledging that there are "immoral websites that are not compatible with Islam," he seemed to hedge his bets either way. Further, his arbitrary and shockingly low number of censored websites—"the majority of them are pornographic"—seemed utterly unbelievable to the journalists present.

At the end of the press conference, Memarian walked straight up to President Khatami to pass along information about his NGO, Volunteer Actors.[64]

"Mr. President, this is a copy of our magazine," he said politely in Persian.

The Iranian bodyguards pushed him back. Memarian stood his ground, annoyed at how even on foreign soil these men, who operated under Khatami's authority, continued to be rough with the public.

"Mr. President, this isn't Qom—this is Geneva!" he retorted back at Khatami, comparing the openness of this European city to the seat of Iran's religious schools.

Khatami immediately looked straight at Memarian and stopped. Reaching across the burly torsos of his guards, he put his hand on Memarian's shoulder.

"Don't I know you?"

"President Khatami, I am Omid Memarian. I am a journalist."

"Ah, yes, how are you? Where are you writing now?"

The president pulled him inside of his security circle and the two chatted amicably for a few minutes as the entourage made its way toward the exit. Memarian reminded him that they had met previously in Tehran along with other journalists, and Khatami apologized for his guards. Of all the other press and foreign dignitaries that gathered around the president, Memarian was the only one whom he singled out.

While Memarian had a chance to chat with the president, following the press conference, the four reporters from *Daily Summit* interviewed Ahmad Motamedi, Iran's newly sworn-in minister for Information and Communications Technology, who confirmed the figure of 240 blocked sites, but still gave confusing and baffling answers:

CARA SWIFT: Will you officially publish the list of the 240 sites that are banned?

MOTAMEDI: Actually it has been published for the private sector. Most of them are private, and all the press know what it is.

AARON SCULLION: What punishments can people expect if they publish websites you do not agree with?

MOTAMEDI: Only we cut the sites—from only access from Iran. There is no punishment defined for them.

SCULLION: There were reports a few months ago that one weblogger was arrested.

MOTAMEDI: No one has been arrested. If you have any name we can follow it. We give some loans to them and promote these weblogs and sites when they are good—especially when they are in Persian.

DAVID STEVEN: I have the name of the weblogger that it was claimed was arrested. Sina Motallebi. [sic]

MOTAMEDI: Actually, it is just now that I am hearing this from you. This is not substantial and it is not in relation to weblogs. What news agency?

STEVEN: The Columbia Journalism Review. Associated Press.

MOTAMEDI: [He] has been arrested but not in relation to weblogs. If somebody is a weblog writer, and kills somebody—should they not be arrested?

SCULLION: Previously, some web sites in Iran were taken off line and blocked— when the government was told of this, it said it was a mistake. How could such a mistake happen?

MOTAMEDI: Technical problems always happen. But I don't know how this is being increased. Sometimes mistakes happen.[65]

Indeed, sometimes mistakes happen.

DURING THE LATTER HALF of 2003, a UN envoy to Tehran met with many political cal dissidents, who told him of illegal imprisonment and torture. Finally, on April 28, 2004, the Iranian government issued a statement attempting to "modernize" its justice system, saying that it would no longer tolerate torture as a legitimate interrogation technique. Further, it would ban the practice of blindfolding detainees and concealing the location of their detention.[66]

While this out-of-character shift took many Iran-watchers by surprise, just six months later Iran was back to its old ways. On September 24, 2004, Hossein Derakhshan wrote on his blog about an article titled "CIA Runs 'Spider's Web' in Iran" in *Kayhan,* the hard-line conservative propaganda newspaper. In an editorial, Editor-in-Chief Hossein Shariatmadari "exposed" a vast network of "Internet journalists and bloggers, inside and outside Iran, who have shaped a CIA-led, sophisticated network in order to undermine the Islamic Republic of Iran and organize large attacks against it."

Derakhshan wrote:

The most alarming part of the piece is where it has named many younger Iranian journalists (with their initial first name and full surname) who still live in Iran, including the recently arrested journalist/blogger/

technicians such as Babak Ghafoori Azar, Hanif Mazrooie, Shahram Rafizadeh, and Roozbeh Mir Ebrahmi among many others.

Based on previous experiences, Kayhan always illustrates the whole picture after each of these scenarios get started by several arrests. So we all should be worried about the fate of the young innocent journalists that, probably just for bad luck, have been fitted into this desperate scenario that tries to find the CIA's hand behind the entire politically active part of the Persian Internet.

What are the implications? First, it proves, at least to me, that our recent protest has been so effective that we have made them react this desperately and harshly. Second, it shows the fact that hardliner conservatives see Internet as a threat to their interests and therefore act against it, proves it as a potentially powerful medium for promoting democracy and freedom of expression which deserves more attention from the Western countries and media. Third, it displays that the number of Internet users in Iran (between 5 to 7 million) is big enough to worry conservatives about its influence.[67]

Imprisoned (2004)

Tehran

October 10, 2004

When the van finally came to a sudden stop, Memarian's curled body thrashed against the back of the seats.

"We're here," the driver said flatly.

The door of the van sprang open, and sunlight warmed Memarian's blindfold as he struggled to get to his feet and regain his balance. He groped for the seat and gripped an unforgiving hand. Stepping outside onto solid pavement, he could hear birds chirping, but the men grabbed him by the upper arm and pushed him forward.

"Go on," one guard grunted at him.

Memarian walked slowly, trying to get his bearings, trying to adjust to this new environment, and most of all, trying to not to provoke a beating. They marched him through a ground-level doorway and down a quiet hall, where the light got dimmer as they walked. Suddenly, the guards grabbed both of his shoulders and turned him 90 degrees. Still blindfolded, the tip of his nose grazed a semi-coarse wooden wall.

He was forced to surrender his belt, wallet, and shoes and sign a piece of paper acknowledging that they had been taken and inventoried, still unable to see the piece of paper that he had signed. He stood for what seemed like an eternity, straining to hear the voices of gruff men down the hallway. Memarian could hear a new set of footsteps approach and turned his head toward this oncoming sound. Suddenly he felt a pair of hands against his torso and was shoved into the wooden wall.

"Didn't I tell you not to move?" the man spat at him. "What did I tell you? Do you not understand the words I'm saying?"

The wall gave way a little bit and Memarian realized that he was standing in front of a door. The guards flanking him exchanged a few more words and opened the door, pushing him inside. There, they finally removed the blindfold and Memarian could actually see the state of his predicament.

The room was barely five paces long and only a pace and a half wide. No natural light entered—and it didn't help that the walls were a thick, dank green color. A light odor of urine pervaded the room.

As Memarian took in the size of his cell, the guards curtly told him that if they needed him, they would bang on the door and he would have to put on his blindfold. They also told him that he should address each of them, and any other guard, as "*Seyed*," an honorific title reserved for those who are descendants of the Prophet Mohammad. Further, he should address the interrogator as "*Haji Agha*," or "Mister Haji," a prestigious name for a Muslim man who has been on a pilgrimage (*hajj*) to Mecca. (This interrogator later would also have the detainees sometimes address him as *Keshavarz*, or Farmer.[68]) Before he had a chance to even process what they had said, the door clanged shut behind him and the sound of the lock rattled for a fleeting moment.

Some light came into the room above the door as Memarian turned his head back toward where the men had left him. On the concrete floor were two disheveled blankets and nothing else. Not even, quite literally, a pot to piss in. In just a few short hours, Memarian had gone from working in an office to standing in a dark, stinky cell. Her had given up his watch, so he couldn't know for sure how many hours had passed.

Here, no one could hear him scream—no one that mattered, anyway. Not a single person on the outside knew where he was. He didn't know where he was. The implication of being in an unmarked, secret prison was clear. There were no rules, no appeals, no phone calls, and certainly no Geneva Convention. That was entirely the point.[69]

In fact, within days following Memarian's arrest, his mother, Touran Memarian, wrote a letter to President Khatami asking where her son was. A few days later, the president replied, saying he didn't know, but that he would look into it.[70]

After what seemed like an hour, a bang came against the door. Memarian put on his blindfold, not wanting to find out the punishment for resisting. The guards took him by each arm and walked him, shoeless, down the length of the hallway. Later, he realized that his cell was one of five, and that there was a matching set of five opposite them. At the end of the short hallway were a restroom and three interrogation rooms, one of which Memarian was shoved into. He felt his body slide into a cold, metal chair that nearly made him jump when he touched it. The bright space was nearly triple the size of his cell.

There, waiting for him, was Haji Agha, a stout and dour man in his mid-fifties, who sported a thin, but full beard. He was seated in an equally simple metal chair on the opposite site of the plain wooden table with a thick dossier, a stack of blank paper, and a pencil. The guards left after removing Memarian's blindfold. No sooner had the door closed than Haji Agha rose from his chair and approached Memarian.

Without saying a word, his hand swept through the air to strike Memarian's cheek. The sharp sound pierced the room. Memarian nearly lost his balance in the chair and his arms flailed as he attempted to remain upright.

"What the hell have you done?!" Haji Agha demanded.

Memarian gasped for words, but none came out.

"You'll stay here for six months, maybe a year!"

Memarian struggled to catch his breath while Haji Agha stormed around the room, which was covered from wall to ceiling, and around the doorframe, in soundproofing foam. Memarian remained silent.

Haji Agha walked back to his side of the table and shoved the stack of nearly fifty pages toward Memarian and rolled a cheap pen along with them.

"Ultimately," he began, "you'll stay, you'll stay, you'll stay *here* until you say everything you have to say. You're going to tell us about everything you've fucked with. Write it down."

Haji Agha's tone quickly softened, saying that they just wanted Memarian to be "loyal and straightforward" with them and that based on what he would write—the entirety of his activities with NGOs, journalism, and blogging—the Judiciary would be able to determine Memarian's honesty. They already had "everything," but wanted to test how cooperative he was going to be.

Haji Agha, with a wave of his hand, dismissed Memarian back to his cell, where he sprawled out on one of the two coarse blankets and began to write. When he lay down, his head nearly touched the door and his feet touched the opposite wall. In this position, with his arms outstretched, he could touch both sides of the room. Memarian detailed his activities with NGO organizations, how he became a journalist, and how he became a blogger. He wrote about his trips overseas with those organizations, and the people that he met and interviewed while a journalist.

After two hours of writing, he was again summoned to see Haji Agha. As Memarian entered the room, Haji Agha didn't even look up from the dossier that he was flipping through. Sitting down across the cold desk from him, Memarian handed him what he'd written. Writing about his own activities was easy, as most of the work that he'd done had been in the public realm, and many people knew about it. It wasn't a secret. Keeping the attention focused on himself meant keeping his friends safe.

Haji Agha took one look at the papers, pawed through a few of the sheets in the middle and end of the stack. He looked up, took one glance at Memarian, and rapidly tore all the pages, once vertically, watching them fall to the floor like oversized confetti.

"This is bullshit," he proclaimed. "Go and write the truth, with details."

He handed Memarian a fresh stack of blank, unlined pages and dismissed him again.

This continued for hours, three or perhaps four times. Memarian lost count. His whole world had been stripped away and reduced to his tiny cell, the march to the interrogation room, and Haji Agha. By the time Haji Agha let him go to sleep, he guessed that it was 2 or 3 A.M.

"Come and get ready, we're going to take you to the bathroom!" yelled one of the Seyeds, banging on Memarian's door some hours later. He hadn't slept well—he was cold most of the night, as he'd used one of the blankets as a pillow and put the second one underneath his body as a sheet. In this secret prison run by the Islamic Republic, all the prisoners would pray. The fact that most reformist-minded writers, like Memarian, were only nominally Muslim was immaterial.

Before any Muslim prays, he or she is expected to cleanse the body, inside and out, of all things unclean. That means using the toilet and washing one's hands and feet for ablutions. Memarian reached for his blindfold, put it over his eyes, and waited for the door to open. Once the pair of Seyeds arrived, they marched him down the hallway again and told him that he could take off the blindfold once he entered the bathroom and that he would only have a few minutes to wash.

Upon entering the bathroom, Memarian removed his blindfold and was shocked to see three other journalists that he was acquainted with: Hanif Mazrooie, Shahram Rafizadeh, and Roozbeh Mirebrahimi. The trio weren't surprised to see Memarian. His name had come up a few times during their interrogations. They had heard someone being brought into the prison the day before, but they didn't know who it was.

Mirebrahimi had been held for twelve days, Mazrooie for thirty days, and Rafizadeh for forty.

As they washed together and used the doorless toilet stalls, the three gave Memarian some tips on how best to survive this decrepit situation. They couldn't look directly at him while they talked, as there were video cameras mounted above the toilet stalls and the guards directly outside.

Mirebrahimi spoke first, speaking quietly while brushing his teeth.

First, he told him, write big, so that it seems like you're writing more than you actually are. Second, don't mention any names. They will arrest and intimidate those people and perpetuate the cycle of harassment and imprisonment. If Memarian was forced to name individuals, Mirebrahimi instructed him to use the names of a high-ranking politician or someone who had left Iran for good, as those people were essentially untouchable.

Third, don't drink much water, to minimize the need to urinate. Memarian, like the other prisoners, would only have three occasions to go to the bathroom in a twenty-four-hour period. Those three occasions would be timed with each of the three main daily Islamic prayers (out of the five daily prayers). Fourth, don't eat a lot, because you'll throw up during the beatings. Fifth, keep one of the food containers in your cell, rinse it out during the prayer breaks, and use it if you absolutely need to use the restroom and don't know when the next break will be. Sixth, play with words and use the writing time to provide more analysis than actual, useful information. Seventh, don't show sensitivity to anyone, including your family. If you insist that they don't harm or harass someone, they'll know that you're vulnerable to that threat.

And with that, Mirebrahimi put back on his blindfold and headed out. The others followed suit. It all happened so fast, Memarian thought—he barely had time to respond, or to tell them how his interrogation went, or anything like that. But there would be two more prayers later on that day, and countless more to come.

The following day, the interrogations continued. The cycle never abated, with Memarian going to see Haji Agha, his ordering Memarian to write about his activities, returning, questioning, and going back to write more. Each time, Memarian would be lucky if he came away at the end of the day with just a few beatings—sometimes he would get punched, kicked, or whipped with Haji Agha's thick leather belt.

Although Memarian didn't know it at the time, that same day, the second day after his arrest, the head of the Judiciary, Ayatollah Mahmoud Hashemi Shahroudi, proclaimed new laws specifically addressing "cyber crimes." The BBC reported: "According to the new law, 'anyone who disseminates information aimed at disturbing the public mind through computer systems or telecommunications . . . would be punished in accordance with the crime of disseminating lies.'"[71] The laws were a clear aim at Memarian and others being held at the time in secret detention.

BY THE END OF THE FIRST WEEK, the interrogations turned to Memarian's blog. As Haji Agha continued his questioning, Memarian began to realize that Haji Agha really didn't know much about what blogs were, or even the basics of the World Wide Web. Many of Memarian's articles and blog posts were either linked to, or republished on, a popular Europe-based Persian-language site called Gooya, which was and is blocked from within Iran. Haji Agha accused Memarian of collaborating with the enemy, essentially, based solely on the fact that they had linked to some of his posts.

Memarian, in turn, explained that linking was not something that he had control over. It was a simple component of the web—in fact, its main innovation—that one could "link" different web pages together. Memarian knew that anyone who had spent five minutes online understood this simple truth. However, Haji Agha did not, and could not. He simply was unable to grasp the concept that Memarian's work could be mentioned or quoted on the site without a direct and explicit order from Memarian.

Haji Agha accused Memarian of having a regular relationship with Farshad Bayan, the founder of Gooya. Indeed, Memarian had met with Bayan at the United Nations World Summit on the Information Society in Geneva in 2003 in an interview for the reformist newspaper *Yas-e Noh.* In reality, Memarian's interview was a bit critical of Bayan and Gooya. But that didn't matter in the eyes of Haji Agha. All that mattered was that Bayan and Memarian had met, and that Gooya linked to Memarian's work, and therefore, there was an "ironclad" case to be made for treason.

In the end, Memarian admitted to "faxing" his articles to Gooya—an answer that Haji Agha accepted. While this answer was far from the truth, it seemed that telling the truth was only going to keep him in prison longer. Haji Agha went on, asking about some posts that Memarian had written, citing printouts that he had in the dossier.

In the weeks before his arrest, Memarian had told a friend that, should he be arrested, the friend should delete key words and phrases but not to delete entire posts. In so doing, Memarian might be able to mitigate his "guilt," and yet make it not apparent on the blog that he was trying to conceal his blogging. Clearly, Haji Agha was going after a blogger because he'd been ordered to—he'd been informed that this man, Memarian, was an enemy of the Islamic Republic. But to him, a blog might as well have been written in stone—he was unable to conceive of the idea that one's printed words online could be modified.

Memarian tried to explain that just because Haji Agha held printouts of some posts in the dossier didn't mean that they were permanently stored online in the same form. He added that online articles can easily be changed, and in a world of self-publishing, the blogger can easily edit, modify, or delete things at will.

"A blog is a self-publishing thing," Memarian pleaded. "I'm sure that I deleted these things."

But Haji Agha wouldn't hear of it. He had the printouts that proved Memarian's propensity to insult Islam and the Islamic Republic, the posts talking about the government's inability to adequately address the needs of its younger citizens. He went on, asking for a "map of the blogosphere," and asked Memarian to provide a list of the ten most influential Persian blogs in literature, and also in politics. Clearly, someone had instructed Haji Agha to ask about blogging, but he was ill-equipped to comprehend how it worked. The questions continued about how enemies of the Islamic Republic can use Persian blogs to gather intelligence, and Memarian couldn't give an answer that Haji Agha would accept.

The truth was that there was no blogging conspiracy, no plot to use blogs to take down the government. Memarian tried to explain this to Haji Agha, but his interrogator didn't understand.

"To answer these questions, it would be good to have a conference," Memarian said. "The bloggers will tell you all this stuff—there's no need to arrest us."

During one interrogation session, Haji Agha began straightaway without any of the minor pleasantries that Memarian had grown accustomed to.

"How did you get your visa?" he stated, his voice rising slightly. He stared down Memarian, who sat with his legs closed and his body braced.

"What did you do in the U.S.?" Haji Agha said, rising from his chair while flipping through the dossier. He tossed the dossier back down onto the table and began walking toward Memarian.

"We have all your information about your meetings in Washington, your meetings with royalists." He stared at Memarian again, fuming.

Memarian was dumbfounded—royalists? What did the family and descendants of the deposed shah of Iran who lived outside Washington, D.C., have anything to do with him?

Haji Agha was referring to the time in 2003 when Memarian had been scheduled to fly to the United States to attend a conference. His flight took him from Tehran to Frankfurt, where he was to catch a connecting flight to Washington, D.C. However, for some inexplicable reason his name was on the United States' no-fly list, and he was sent back to Tehran. He never made it past Frankfurt airport.

The ignorance about the Internet and blogging was one thing, but this was the first time that Haji Agha had made a mistake. It was also the first time that Memarian could clearly see his bluff.

"I wasn't in the U.S.," Memarian said flatly, staring at the nearly empty desk. "My name was on the no-fly list. If they'd wanted me to go, they never would have put me on the no-fly list."

Haji Agha's eyes flared while his body stiffened. Each step of his black leather shoes resounded. Before Memarian had a chance to see Haji Agha's fist approach his cheek, he was sprawled out onto the floor.

"Liar!" he yelled, nearly spitting at Memarian. "Do not fuck with me!"

He kicked Memarian repeatedly in the stomach with his leather boot. Resisting, Memarian covered his face and curled into a fetal position, but the swift kicks would not abate. His torso absorbed most of the blows. He tried to remember what Mazrooie had told him one day during a bathroom break—that the beatings would always be short, less than ten minutes. All he had to do was just sustain the blows. But this time, as Memarian got thrashed around, the dust began to rise up from the thin wall-to-wall carpet as his gut twisted with each kick.

"Why are you lying to me?" Haji Agha screamed at him, kneeling down into Memarian's dusty, teary face. As Haji Agha rose and delivered another heavy pound of his shoe, Memarian turned his head and began to vomit. The Haji Agha hopped away, so that he wouldn't step in it, but was clearly terrified at the reaction. Memarian continued his convulsions and summoned every raw emotion into the most guttural sound he could muster. He didn't want to be hurt more, but he thought grunting and aching would at least abate the interrogation temporarily.

Haji Agha reached immediately for the door and cracked it a few inches.

"Guards!" he bellowed.

Memarian continued his moaning and was sure that the other prisoners could hear him now. As he tried to sit up in a room filled with dust, vomit, dirt, and distress, a pair of Seyeds came and hoisted him to his feet. Without saying a word, they took him to the bathroom so he could wash out his mouth and clean his face. After a few minutes, they led him back to his cell.

One reached into his shirt pocket and produced a sealed capsule.

"Here, this will help you sleep," he said, his eyes locked in sympathy with Memarian's.

THE MONTH BEFORE MEMARIAN was arrested, a handful of his colleagues and friends had also been sent to the same prison on equally dubious charges. In fact, Memarian was the last of a group of twenty-one bloggers to be arrested, held in solitary confinement, and tortured by the Islamic Republic during September and October 2004.

Among the earliest arrested was Shahram Rafizadeh, a twenty-nine-year-old journalist who had written for *Shargh, Khordad, Azad* and other journals. At the time, he was the political editor of *Etemaat*. He, too, was arrested for writing in his weblog about a few Iranian writers who had previously been arrested as political prisoners. A pair of men, who didn't identify themselves, came to his office late in the afternoon on September 7, 2004. Initially the men told him that they "just wanted to ask him a few questions," and that it would only take a short time. Rafizadeh suspected that he was being arrested, and so he questioned his captors directly.

"I asked them, 'If you want to come and take me, then tell me,'" he recalled.[72] "'If you're going to take me for a couple of months or years, tell me now, so I can know—so I can make a phone call to my wife and my children.'"

Just as with Memarian, they took Rafizadeh in the back of an unmarked van with tinted windows and curtains and drove him home, where they confiscated nearly all of his books, papers, and photographs. After strip-searching him and putting him in a dark and minuscule cell, he was brought before the same Haji Agha, and immediately was beaten to the point of unconsciousness. He awoke some period of time later with his head under the bathroom sink, where the guards were trying to revive him. The Seyeds dragged him back to Haji Agha, who Rafizadeh recalled as saying immediately after his first beating: "This is just the beginning and I'm going to take the skin off of your skull and I'm going to kill you and I'll make your cell your grave."

Rafizadeh found that next door to his cell was Saeed Motalebi, the elderly father of Sina Motalebi, who had been arrested shortly after his son had fled to the Netherlands. The following day, journalists and bloggers Hanif Mazrooie and Babak Ghafoori Azar both were arrested and brought to the same prison and suffered the same treatment.[73]

Ghafoori Azar wrote for *Cappuccino,* and following his arrest the editors voted to permanently close down the magazine, as one of his "crimes" was having written for the online publication. As most of the editors remained in Iran, they, too, were afraid of being targeted. As of 2008, the final edition from September 2004 sat on the homepage of cappuccinomag.com, frozen in time. There was no explanation anywhere on the site as to why they stopped publishing, nor what happened to its editors and writers. It lived on as a virtual Pompeii, until sometime in 2009, when the server was suddenly taken offline.

On September 27, 2004, Roozbeh Mirebrahimi, a twenty-five-year-old blogger and journalist, was also brought to the same prison. He was the political editor of *Sharvand,* another reformist newspaper. He was arrested on a late summer morning on the south side of Tehran, where he shared an apartment with his wife and with Rafizadeh and his family.

When Memarian was brought in, his cell was right next to Mirebrahimi's. Like Memarian, he experienced multiple beatings and direct threats against his friends and family members at the hands of Haji Agha. He was accused of undermining national security, publishing lies, and writing libelous statements against Supreme Leader Ayatollah Ali Khamenei, among other charges.

IN LATE NOVEMBER, following pressure put on the Iranian government by Reporters without Borders, Human Rights Watch, and other groups, many of the other prisoners—mainly webmasters and other technicians—were released. However, Memarian, Rafizadeh, Mazrooie and Mirebrahimi were transferred to Evin Prison, an infamous facility. Evin's Section 209 held the highest-profile political prisoners in Iran, including Akbar Ganji, the famed democracy activist and writer now living in Canada. At the same time that Memarian and the others were brought to Evin, Ganji was serving a six-year sentence.

There, Memarian, Rafizadeh, Mazrooie, and Mirebrahimi were all fingerprinted, photographed, and even had visitation rights. While in prison, Memarian and his fellow prisoners continued to be interrogated and met with other prisoners who had been there for years at a time and who were undergoing much harsher interrogations than they were. After several weeks in Evin, Memarian could feel his personality turning harsher. And, having watched other prisoners around him, he knew that the longer he and his friends remained, the lesser the chance they would be able to leave.

"Whatever they want us to say, let's do it outside [in public, out of prison], because no one will hear us here," he recalled saying later.[74] "I had a feeling that I was getting to an irreversible point. I thought that psychologically if I stay here for one more month or two more months that I can't come back to my normal life. [I thought that] we should try to leave here and do whatever they want and go say whatever they want us to say." Besides, Memarian and the others knew that whatever it was that they confessed to, no one would believe it. Their families and friends would know that they were forced to write and say whatever it was—and regardless, they would simply be overjoyed to have their sons home once again. They all agreed to write confession letters—Memarian stated that he was "misused by reformist politicians and journalists" and that he was trying to "make a black picture" of the Islamic Republic.

Release and Aftermath (2004–2009)

On December 2, 2004, Omid Memarian, Shahram Rafizadeh, and Hanif Mazrooie were released from Evin Prison. Roozbeh Mirebrahimi had gotten out first, just two days earlier, having been forced to write a similar confession letter. Within just a few days of being let go, Memarian was contacted by Haji Agha, his former interrogator, who told him that the two of them had to go to the police station to file a complaint against some of the reformist politicians who were named in their confessions. Presumably, Haji Agha and his associates would go after these reformists much in the same way that they had the bloggers. However, the reformists whom Memarian had named in his confession were people he'd never met—he had simply named them to satisfy the demands of his jailers. "They started putting pressure on us more and more," he said.[75] "We understood that this game is not over."

Facing the organizers of his imprisonment was no easy task, Memarian recalled later:

The people that arrest us, they know what they are doing. They know that they are putting us in a situation of constant fear and terror. That's a horrible situation. People have no idea what that means. It's like paranoia—if you've seen A Beautiful Mind, the person, the genius guy, [it's like that]—you see people all around yourself. It's less now, but I used to see my interrogators around myself. That's the horrible thing that you can have in your life. I used to see them and they would be trying to intimidate me. I can hear the sound of the solitary confinement's door. I can hear the slapping of my [own] face.

It was just horrible. These are the feelings that stay with you. I think they knew what they [were] doing. This is what they call white torture. A torture that never stays on your body, that doesn't mark on your body but it stays with you. It never leaves you alone.[76]

Memarian enlisted the help of former Vice President Mohammad Ali Abtahi, who served in the Khatami government. Abtahi himself was a well-known blogger, and had resigned less than a week before Memarian's arrest, saying that he could not work with the hard-line conservative Parliament.[77] Abtahi was able to exert his influence over President Khatami to create a presidential commission of inquiry into the detentions of Memarian and others. Abtahi served as the official presidential representative on this commission.

In December 2004 and January 2005, the commission heard unprecedented testimony from Memarian, Rafizadeh, and other bloggers who were subjected to imprisonment and torture. After hearing their testimony, Abtahi wrote on his own blog that these men had, in fact, been tortured. He also wrote that during the January 1, 2005, commission meeting, where Memarian and Rafizadeh testified, their statements were so shocking and emotionally charged that "we had to give them water so that they could get hold of themselves and continue."[78]

Saeed Mortazavi, the chief prosecutor who was ultimately responsible for the bloggers' detention, had previously warned the pair not to speak of their experience in prison. While Abtahi pushed the limits by writing about this commission, President Khatami and the head of the Iranian Judiciary, Ayatollah Mahmoud Shahroudi, both promised to follow up on the accusations made by the bloggers.[79] Shahroudi later ordered an internal investigation, but ultimately no government official involved in their detention was ever brought to account.

Abtahi's attention on the case eventually enabled local Iranian newspapers to begin covering the cases of the arrested bloggers, which in turn drew the attention of the larger international journalism, blogging, and human rights communities.[80]

Following these hearings, a large number of the pioneers of reformist journalism and blogging left the country. Salman Jariri, the first Persian blogger, emigrated to Dubai for work. He now tends to shy away from politics and prefers to have a low profile. Even today, his blog is still free of comments—in contrast to the way nearly all other blogs are set up.

Sina Motalebi received political asylum in the Netherlands and now works for BBC Persian in London. Roozbeh Mirebrahimi has been living in New York City since 2007. Shahram Rafizadeh moved to Toronto in early 2008 and continues to report for expatriate Persian-language media. A large number of the original group of editors of *Cappuccino,* including Sanam Dolatshahi, have moved on to the United Kingdom, France, and California. Masoud Safiri, the former editor of *Hayat-e Noh,* now lives in Washington, D.C. Hossein Derakhshan briefly attended graduate school in London in 2007. He later returned to Iran in the fall of 2008, whereupon he was arrested in Tehran in November 2008. He remains in prison and has had little contact with his family since his arrest.

Omid Memarian left Iran for the United States in 2005, and was awarded a "Defender of Human Rights Award" by Human Rights Watch.[81] He was later given a scholarship in order to attend the University of California, Berkeley, where he pursued master's degrees in both peace and conflict studies and journalism. He graduated in May 2009, and after working for Human Rights Watch for six months following his graduation, he has gone back to journalism.

Mohammad Ali Abtahi, once a respected member of the Khatami government, and an ally of Memarian's, was arrested in the aftermath of the June 2009 presidential election and was released on bail in November 2009. Ali Reza Eshraghi joined Memarian in the United States and became a visiting scholar to the University of California, Berkeley, in 2008, took a journalism job in 2010 in Washington, D.C., and began graduate work at Duke University in the fall of 2010.

The exodus of a large portion of Iran's younger generation of writers and thinkers echoes the similar journeys of previous generations of Iranian intellectuals who were driven out following the Islamic Revolution. Many journalists who fled the revolution set up Persian-language television stations in Europe and the United States. Today, those television channels are viewed with a great deal of skepticism inside Iran—their backers have largely remained outside Iran for nearly three decades.

Part of the magic during the heyday of Iranian blogs of the early 2000s was that they had come from within Iran. Writers, amateur and professional alike, pushed the limits of online speech from within Iran in ways that had not been possible previously. They took advantage of technologies that made it easier for them to type, consume media, and publish in their own language. The most influential blogs all started out from within Iran, and yet their authors have largely emigrated elsewhere. While many of them say that they would like to return to Iran, to date, very few have done so. Indeed, as the longtime Tehran resident and British journalist Christopher de Bellaigue notes in his book *The Struggle for Iran:* "Exiles are able, as Iranians in Iran are not, to tell the full truth about the Islamic Republic. However, the longer they spend away from Iran, the less acquainted they are with it, and the more their accounts are open to question."[82]

As such, while this mass migration of a large group of the "Children of the Revolution" takes root, one can't help but worry that the longer they stays outside Iran, too afraid to face the wrath of the Islamic Republic at home, the greater the risk that they will be permanently expatriated.

Some bloggers like Memarian argue that as Iran's level of Internet connectivity increases, particularly among the younger generation, that even those outside the country can stay tapped into what is going on in ways that were not possible before. Today, from his apartment, Memarian monitors Persian-language sites such as BBC Persian, Gooya, Rooz, and many blogs that are written inside the country. He is in regular contact with many bloggers and politicians, and can easily contact any number of his sources inside Iran with a quick instant message or phone call.

> We have this chance—the Internet is a huge factor that lets us be connected and make the best use of that, to be in touch with the realities, different aspects of society. The [older generation of] people who went to L.A. [after the revolution] are not the people who are good in that, in that kind of communications. They have no connection to the society. The people who were active in the revolution are not active now. The society has entirely changed. They don't have a connection with the changed society. We were there. Still now, I can call and talk to a member of Parliament. If something happens, I can get the number of somebody who has done something or has been in the center of an event, [I] can get access to people very quickly.[83]

New York Times reporter Nazila Fathi echoed this sentiment in an article written in January 2010. She describes how, after being forced to leave Iran in the aftermath of the 2009 presidential election, she was afraid that she would lose contact with her homeland.

> For me, that was like a new dawn: rather than being cut off, I had made contact with another Iran—a virtual one on the Internet, linking reformers abroad to bloggers and demonstrators still inside the country, and to reporters and sources outside. In fact, by following blogs and the cellphone videos seeping out of Iran, in some ways I could report more productively than when I had to fear and outwit the government.
>
> For example, my contacts helped me find and interview a young man who had left Iran after being in prison, where he said a guard had raped him. That interview could not have happened in Iran. Last month, I could freely translate the harsh slogans that protestors hurled about the supreme religious leader, Ayatollah Ali Khamenei. There were palpably genuine videos on YouTube from places I recognized, with crowds chanting slogans I knew—or new ones. The slogans were now in fact fiercer, the leaders of the movement less timid, and at least some of the demonstrators clearly angrier.
>
> So I could report, free of government edicts, that the protests were entering a new phase, even as I remembered a cardinal self-imposed rule for any reporting from Iran: There is no way to predict where any movement might be heading, or when it might be stopped.
>
> There is an irony in all this; the years of authoritarian control had educated much of Iran in the need for circumventing restrictions on the Internet, and now I was seeing and hearing the results on my computer and television.[84]

While it is not entirely surprising that many reformist intellectuals and journalists are writing online—ranging from former Vice President Mohammad Ali Abtahi to investigative journalist Massoud Behnoud to someone like Omid Memarian—it is a little bit unexpected that political candidates have set up blogs. During the 2005 election, many presidential candidates had blogs, as a way to reach out to prospective—particularly younger and more secular—voters. Most of those blogs were eventually shuttered or abandoned as their respective candidates were dismissed or lost. In the 2009 election, many candidates, including conservatives, turned to Facebook, Twitter, and other social media to get their message across. In other words, the Islamic Republic itself is starting to catch on, and is at least attempting to co-opt this medium and use it for their own purposes.

IN JUNE 2005, the former mayor of Tehran, Mahmoud Ahmadinejad, was elected president of Iran. He succeeded Mohammad Khatami, the outgoing moderate cleric who led the country for eight years and who, after serving two terms, was barred from running for re-election. He defeated his main rival, former President Ali Akbar Rafsanjani, on a platform of bread-and-butter issues, such as the price of meat, onions, and gasoline.[85] While the Tehran-based political establishment and generally more moderate urban residents initially dismissed Ahmadinejad, he drew wide support from Iran's provinces and rural areas. The reaction in the moderate and liberal Iranian blogosphere—which had largely ignored him until he placed second in the first round of voting—was mostly critical of Ahmadinejad's election.

Within six months of his election, he became a household name around the world, with his widely repeated mistranslated line about Israel needing to be "wiped off the map" and his infamous claim that the Holocaust was a "myth." More dangerous to the world community was the fact that during this period Iran also began re-enriching uranium. Within less than one year of his election, President Ahmadinejad announced to the world that Iran had successfully enriched uranium. As the Ahmadinejad presidency moves forward, he has continued to be a thorn in the side of the international community, particularly by aligning himself with other anti-American leaders, such as President Hugo Chavez of Venezuela.[86]

However, one of the most surprising things about Ahmadinejad's presidency is that on August 8, 2006, he himself began a weblog, entitled "Mahmoud Ahmadinejad's Personal Memos." The blog, ostensibly written by the president himself, is translated from the original Persian into French, Arabic, and English. There are also comments in all four languages, an apparent attempt to spread his message far across the globe.

His posts range from the autobiographical:

During the era that nobility was a prestige and living in a city was perfection, I was born in a poor family in a remote village of

Garmsar—approximately 90 kilometer east of Tehran. I was born fifteen
years after Iran was invaded by foreign forces—in August of 1940—and
the time that another puppet, named Mohammad Reza—the son of Reza
Mirpange—was set as a monarch in Iran.[87]

To the celebratory:

Merry Christmas to everyone!
    My sincere congratulations to everyone for the Glorious and Auspi-
cious Birthday of Divine Prophet—confirmed and authenticated by
Gabriel, the angel of Divine revelation—the Obedient of Almighty God,
    Jesus Christ, the Messiah (peace be upon Him)
    He was a messenger of peace, devotion and love based upon
monotheism and justice. He was raised in His Mother's hand—Virgin
Mary (peace be upon her)—that Almighty God stood her as impeccable
and exalted her above the women of the world. The Mother and the Son
that in the Divine Sight are reputable and prestigious. And they are
positioned by God—The All Wise—at a sublime level.[88]

And finally, to the academic:

One's perspective regarding government and governance determines the
way one should cooperate with the people. If one recognizes government
as a privilege and prey of the governors, then the period of governance can
be counted as an opportunity to fulfill the expectations of certain individuals and groups or the ostentation and hedonism of the governors.[89]

Following an unexplained eight-month silence that ended in late 2007, his
blog drew renewed attention from international media.[90] His resumption of
blogging indicates that someone in the highest levels of the Iranian government (perhaps even the president himself) had believed in the importance of
using blogging to further the Islamic Republic's message. However the blog
went silent again in 2008.

Conclusion (2009–Present)

Over the years, the number of Persian blogs has grown from several dozen to
thousands. Yet it remains unclear who the winners and losers are in this interplay between the hard-line government and many in the moderate, reformist,
and secular blogging public. To be fair, there are many conservative and
Islamist bloggers, as Hamid Tehrani, the Iran editor of globalvoicesonline.org,
points out in an article from late 2007:

In the last two years, Islamist bloggers became much more active and
organized than before. Mahmoud Ahmadinejad's victory played a key

role in mobilizing these blogs in different ways. Reformist bloggers found themselves out of power and started to use the blogs as instruments to get votes. Government itself supports—directly or indirectly—organizations such as the Office for Religious Blogs Development (ORBD). This office has a project to help every religious student get a blog.[91]

Still, many Iranian bloggers seem to agree that a significant portion of their fellow bloggers tend to be better educated, urban, and thus more likely to be moderate and secular. However, as the Internet continues to grow rapidly in Iran, it is only natural that the government and its ideological allies will use the medium to spread their own message as well. While bloggers like Omid Memarian and his contemporaries continue to publish, the government has been successful in using intimidation to drive many of them out of the country.

The bloggers who have left Iran could be viewed as being the winners in this game, as they have broken free of the shackles of the Islamic Republic; from the comforts of California or the United Kingdom, they are free to say whatever they want. They are able to express their ideas to their countrymen unhindered by the threat of arrest, harassment, or violence. Many have been able to pursue degrees in higher education, or develop careers as journalists or with NGOs in the West. This is precisely what many want for themselves. There is hardly a twenty-year-old Iranian who wouldn't want to emigrate to Europe, Canada, or the United States for the simple reason that the Iranian economy has been in a deep slump, and there simply aren't enough jobs—let alone jobs for young journalists and writers.

That being said, it is precisely because of the fact that these Iranians, like many of their predecessors in decades previous, are now outside of the country that the government easily dismisses them. Many of those who have been arrested or targeted by the regime continue to be very cautious, and tend to keep a low profile concerning their professional activities. Many Iranians born after 1979 who have left Iran for political reasons over the last several years have family members back home, and there is a lingering thought that hangs in many of their minds that something may happen to them.

In fact, sometimes legal proceedings can effectively make it impossible for some Iranians like Omid Memarian to return home. On February 4, 2009, Iranian authorities convicted Memarian, as well as the other three who were arrested and tried with him—Shahram Rafizadeh, Roozbeh Mirebrahimi, and a fourth journalist and blogger, Javad Gholam Tamayomi—of the crimes of "participating in the establishment of illegal organizations," "membership in illegal organizations," "disseminating lies," and "disturbing public order." Gholam Tamayomi, the only one of the four who still lives in Iran, was also charged with treason. Each was sentenced to prison terms of up to three years and three months, and to be flogged. Memarian was fined 500,000 tomans, or around $520.[92]

Memarian said that the four are appealing their case, but given the fact that it took four years for the sentence to come down, that process may take years. Further, this sentence effectively makes it impossible for Memarian, Rafizadeh, and Mirebrahimi to return to Iran in the foreseeable future.

Despite his own departure from Iran, Memarian views the Iranian government as an "absolute loser" in this struggle:

> We are pursing our life. We are pursuing our dreams. Despite the intentions of the Islamic Republic to stop us, to stop us from writing, from painting our imaginations, we are committed to talk about truth, to whatever extent that we can. What they have done, they have not been successful. Their efforts have failed. They are losers in this game. They are absolute losers in this game. They are losing the capital, the human capital of the country. Every day people like us are leaving the country.
>
> The judiciary system has been so corrupted by some people who can do whatever they want to do and they have absolute power to commit crimes and violate the laws. So they are losers, and nobody in the judiciary system can stand up and say this is wrong—you can't do this. And it has not been effective. This is the ultimate point—this has not been effective. They're wasting their time, energy and money of the country. At the end of the day, life goes on, people are [telling the] truth.
>
> They cannot hide anything; they are losing respect, and legitimacy. And after all these countries [like the United States], under the threat of foreign [countries' pressure], more than ever they are making bad decisions. Because nobody is criticizing them inside the country, nobody is telling them what is wrong and what is right. They are depriving themselves of independent voices, people like us, so definitely they are the losers. We are living our life and to whatever extent we can, we are trying to contribute and to be effective in our society.[93]

When Omid Memarian puts his fingers to his keyboard and taps away at a blog post that the Islamic Republic may oppose, while he's using a means of communication that is amplifying his ability to communicate with his fellow citizens, the fact remains that he is but one faraway voice reaching out to a society of millions back home.

While the medium of blogs may provide him and his colleagues, fellow thinkers, and writers with a virtual soapbox on which to speak, the sobering fact remains that the government of Iran controls its legal system. However draconian and Kafkaesque that system may, in fact, be is another issue. But there is a system nevertheless that reserves the right to arrest people that it views as dissidents. Clearly, despite the fact that Iran "no longer tortures people," the government reserves the right to send thugs in unmarked cars to arrest people at their place of work and hold them, essentially incommunicado, in solitary confinement at secret detention centers.

Or as Tim Wu and Jack Goldsmith write in *Who Controls the Internet?*:

> What we have seen, time and time again, is that physical coercion by government—the hallmark of a traditional legal system—remains far more important than anyone expected. This may sound crude and ugly and even depressing. Yet at a fundamental level, it's the most important thing missing from most predictions of where globalization will lead, and the most significant gap in predictions about the future shape of the Internet.[94]

That said, Iran's newest generation of dissident thinkers seem to agree that a revolution that would dismantle the Islamic Republic entirely would be out of the question—this is why they call themselves "reformist." They believe that a true revolution would be both undesirable and unachievable, and therefore seek an evolution—or revision—of the current system, where they would be able to speak freely to one another. This would be a great improvement over what happens now, where they worry about their friends being labeled as dissidents for performing journalism, a public service, to the country that they were born and raised into.

As Memarian himself said in December 2007:

> There's always hope. Five years ago I couldn't think that I would make a decision to leave the country. Three years ago I would never imagine that I would get a chance to get a scholarship. Five years from now, who knows, perhaps I will come back to Iran and run for parliament? Who knows? That's the beauty of life. I'm always open to all options. . . . I think the society has the potential to absorb change. If things happen in a way that tries to put these kinds of things in the right action, people will be ready for that. People are ready for change, if it happens in the right way.[95]

TIMELINE

August 2, 1997: Mohammad Khatami assumes presidency

September 7, 2001: Salman Jariri starts blogging

November 2001: Hossein Derakhshan publishes blogging guide

November 9, 2001: Sanam Dolatshahi starts blogging

June 12, 2002: *Cappuccino* begins publication

August 7, 2002: Omid Memarian starts blogging

December 1, 2002: Sina Motalebi starts blogging

January 8, 2003: Ali Reza Eshraghi publishes 1937 American political cartoon

January 11, 2003: Ali Reza Eshraghi arrested

January 17, 2003: *Hayat-e Noh* shut down

March 3, 2003: Local elections, conservatives win

March 9, 2003: Ali Reza Eshraghi released

April 20, 2003: Sina Motalebi arrested

May 12, 2003: Sina Motalebi released

December 2003: Sina Motalebi flees for Europe

December 11, 2003: WSIS Conference

February 20, 2004: Conservatives take majority in Parliament

April 28, 2004: Ayatollah Mahmoud Hashemi Shahroudi declares: "Any kind of torture of the accused to obtain confessions is banned and confessions extracted through torture will not be religiously or legally legitimate."

September 7, 2004: Shahram Rafizadeh arrested

September 8, 2004: Saeed Motalebi (father of Sina Motalebi) arrested

September 8, 2004: Babak Ghafoori Azar arrested; Hanif Mazrooie arrested

September 10, 2004: *Cappuccino* shut down

September 22, 2004: Babak Ghafoori Azar released

September 24, 2004: "Spider's Web" article published

October 4, 2004: Vice President Mohammad Ali Abtahi resigns

October 10, 2004: Omid Memarian arrested

October 18, 2004: Javad Gholam Tamayomi arrested

October 27, 2004: New law "on punishment of crimes linked to the Internet" announced; Judiciary chief Ayatollah Shahroudi says: " . . . these people will be tried in connection with moral crimes."

November 1, 2004: Mahboubeh Abbasgholizadeh arrested; Mojtaba Saminejad arrested

November 30, 2004: Mahboubeh Abbasgholizadeh released

December 2, 2004: Omid Memarian and Shahram Rafizadeh released

December 25, 2004: Hanif Mazrooie, Massoud Ghoreishi, Fereshteh Ghazi, Arash Naderpour, and Mahboubeh Abbasgholizadeh testify before presidential committee

January 1, 2005: Omid Memarian and Roozbeh Mirebrahimi testify before presidential committee

January 12, 2005: Ayatollah Shahroudi orders internal investigation

April 20, 2005: Abuse confirmed, but Omid Memarian, Shahram Rafizadeh, Javad Gholam Tamayomi, and Roozbeh Mirebrahimi cases are still active

August 3, 2005: Mahmoud Ahmadinejad assumes the presidency

August 8, 2006: President Mahmoud Ahmadinejad starts blogging

November 1, 2008: Hossein Derakhshan arrested

February 4, 2009: Omid Memarian et al. convicted of "participating in the establishment of illegal organizations," "membership in illegal organizations," "disseminating lies," and "disturbing public order." Memarian fined 500,000 tomans, or around $520

March 16, 2009: Omid Reza Mirsayafi dies in custody

June 12, 2009: Iranian presidential elections; Mahmoud Ahmadinejad "wins" reelection

Conclusion

Washington, D.C.

January 21, 2010

On a wintery Thursday morning in the nation's capital, I found myself in the same room with Secretary of State Hillary Clinton. I had been invited to attend the secretary's speech on "Internet Freedom" in a four hundred-person lecture hall inside the Newseum, just blocks from the U.S. Capitol building. Gathered around me were half a dozen senators, dozens of journalists, government employees, and many other Internet activists from Vietnam, Egypt, and Iran. The address was billed as a "major policy address" on the intersection of the Internet and freedom around the world. The address came just one week after Google had announced it would no longer censor its search results in China following a cyberattack on its corporate networking infrastructure. This reversal came five years after Google had decided that it was better to operate a censored search engine in China than no search engine at all.

In a forty-minute address, Secretary Clinton spoke of the Internet in eloquent and intelligent terms in a completely unprecedented manner. In essence, the core message was this: "We stand for a single Internet where all of humanity has equal access to knowledge and ideas. And we recognize that the world's information infrastructure will become what we and others make of it."[1]

The speech was the most public manifestation of what Clinton and other State Department officials have called "Twenty-first-Century Statecraft," the idea of using new information and communications technologies to better shape American foreign policy and advance ideas of freedom of speech, assembly, and religion.

It's no secret that Barack Obama was widely considered to be America's first "Internet president," having built his campaign and his successful election on the backs of social networking websites, micro-donations, and even an iPhone application. As president, he became the first American commander in chief who insisted on using an ultra-secure BlackBerry while in office. His predecessor, George W. Bush, by contrast, famously did not send a single e-mail during his presidency, lest it be subject to later scrutiny or subpoena. However, during the Obama administration, the U.S. Department of State moved quickly to adopt these new tools to better forward his policy agenda of greater global engagement.

Despite the potential of the Internet to harness the power of greater freedom, transparency, and prosperity, Clinton reminded the audience:

> These tools are also being exploited to undermine human progress and political rights. Just as steel can be used to build hospitals or machine guns, or nuclear power can either energize a city or destroy it, modern information networks and the technologies they support can be harnessed for good or for ill. The same networks that help organize movements for freedom also enable al-Qaida to spew hatred and incite violence against the innocent. And technologies with the potential to open up access to government and promote transparency can also be hijacked by governments to crush dissent and deny human rights.

She added moments later that "new technologies do not take sides in the struggle for freedom and progress, but the United States does." In other words, political and economic realities always mold the Internet, and not the other way around. In the United States, where the Internet remains free, the Internet takes on a logical extension of eighteenth-century Enlightenment ideals, where the freedom to assemble on the street and the freedom of the press now translate into the freedom to congregate online and the freedom for anyone to publish anything online.

Clinton's words came just a month after Alec Ross, a high-level State advisor, gave a similar speech at the Brookings Institution. Ross is often given credit for helping to turn the State Department's views on technology around. Like Clinton, Ross pointed out that while Iranian dissidents use the Internet to get information out of the country—such as the tragic murder of Neda Agha-Soltan—the government of Iran is also capable of using the exact same tools to conduct its own propaganda and intimidation campaign.

At that speech, Ross noted:

> While these examples from Iran are compelling to many around the globe, it's important to make clear that just as these networks were used to organize—as well as to galvanize the outside world—they were also monitored and manipulated by government forces. The same openness that allowed sympathizers in, also let in those that sought to end the dissent and punish the dissenters.

So we clearly can't take a sort of kumbaya approach to connection technologies. They can and are being used by our enemies, like al-Qaeda, and by authoritarian regimes. But I think that this, more than anything else, makes the case for our own aggressive engagement on global networks. We need to raise our own game. We can't curl into the fetal position because bad guys are becoming smarter about how to use technology. It just creates an imperative for us to be smarter ourselves.[2]

By using the term "twenty-first century," it is clear that the American government wants to engage very modern communications tools to get its message across. In March 2009, shortly after his inauguration, President Obama recorded a Persian New Year greeting for the Iranian people and made it available, with Persian subtitles, on YouTube. The State Department has tried to make itself more accessible through modern online social networking sites, such as Facebook, Twitter, and Flickr. According to media reports, since March 2009, Secretary Clinton "[has taken] an active role in answering questions from the Web, responding to bloggers and pushing her agency's new media agenda."[3]

THE IDEA OF TWENTY-FIRST-CENTURY STATECRAFT, elaborated Ross in a May 2009 interview, is to move "away from a sole focus on government-to-government interaction and towards government-to-people, people-to-government, and maybe even people-to-people. Government can be much more creative in how they enable people to engage directly with each other," he argues, "and there's no doubt that networked people can become important players on the international stage as well."[4]

For her part, Clinton added in her January 2010 speech that the United States would be working with various partners in and outside of government to "harness the power of connection technologies and apply them to our diplomatic goals. By relying on mobile phones, mapping applications, and other new tools, we can empower citizens and leverage our traditional diplomacy. We can address deficiencies in the current market for innovation."[5]

The speech, which drew a standing ovation, was certainly the most significant American policy statement on the Internet by a high-level government official in recent history. The speech showed that a new, more nuanced narrative of the Internet is beginning to be understood within the highest levels of power in the United States and around the world. This is a view that finally acknowledges the Internet's shortcomings and potential failings, while attempting to champion the best things about it: ease-of-use, transparency, and ability to break down barriers of time and distance. It is notable that the United States is now finally taking a much more sophisticated view of how the Internet can be used for good in the world.

Given that the United States and other world organizations are now attempting to understand the potential role of the Internet in a new way, it is imperative that policymakers and business leaders understand the Internet's global evolution. In order to best engage with twenty-first-century online problems—be they political disagreements or economic development issues—it is crucial to understand the relevant societal aspects, such as education and national drive, that determine how and why the Internet exists within sovereign borders. Government officials need to comprehend that there is logic as to when, why, and how the Internet developed in a particular place. Moreover, there will always be applications, both positive and negative, that reflect the political, economic, and social contexts of any particular nation. Each of the countries that I have described in this book—Estonia, South Korea, Iran, and Senegal and the rest of sub-Saharan Africa—will continue to forge their unique online identities during the coming years on the forefront of innovation, tribulations, and development.

As both South Korea and Estonia have shown, it is quite possible to catapult oneself to modernity through well-orchestrated planning of information and communications infrastructure. Back in the early 1990s, both countries were middle-income nations that were not very significant on the world stage. Thanks to concerted domestic political efforts at home, both countries have done a phenomenal job of harnessing highly educated people to develop highly advanced networks and build new applications like Skype and e-voting. Today, South Korea and Estonia are the epitome of modern, wired (or wireless) societies, and they reached that point largely at their own pace, with their own local flavor.

In contrast to the developed world, in sub-Saharan Africa, Internet penetration still remains low due to bureaucratic incompetence, high cost, and low literacy rates. That's why bringing traditional, computer-based Internet projects to countries like Senegal have failed so miserably time and time again. Meanwhile though, mobile phone ownership has skyrocketed across the continent. That's why Vint Cerf, one of the Internet's pioneers and now an executive at Google, told me in a February 2009 interview that Africa's future may in fact lie in mobile phone use: "So for a lot [of Africans], their first interaction with the Internet won't be on a laptop, but with a mobile. So the things that they're interested in, the products that they can use effectively will be different from the ones that you and I might use with our high-speed laptops."[6]

In Senegal, and other parts of the continent, just as Cerf predicted, new applications designed to connect mobile phones with the web have cropped up. They include: Manobi, a Senegal-based mobile phone tool with which farmers and fishermen can check live market prices by text message; M-Pesa, a Kenya-based mobile phone method of quickly and easily transferring money between people; Project Masiluleke, a South African project, texts reminder messages for HIV/AIDS testing in local languages to mobile phone users. There are countless other examples of how these new types of mobile computing platforms can allow totally new applications for the developing world. Many of these applications, like mobile banking, simply do not exist even in many parts of the developed world, especially in the country that invented the Internet.

Just as the Internet has played a crucial role in vaulting many middle-income countries to upper levels of development, mobile phone technology in many cases can do the same for lower-income countries, bumping them to middle income, as Secretary Clinton said in her speech:

> In many cases, the Internet, mobile phones, and other connection technologies can do for economic growth what the Green Revolution did for agriculture. You can now generate significant yields from very modest inputs. And one World Bank study found that in a typical developing country, a 10 percent increase in the penetration rate for mobile phones led to an almost 1 percent increase in per capita GDP. To just put this into context, for India, that would translate into almost $10 billion a year.
>
> A connection to global information networks is like an on-ramp to modernity. In the early years of these technologies, many believed that they would divide the world between haves and have-nots. But that hasn't happened. There are 4 billion cell phones in use today. Many of them are in the hands of market vendors, rickshaw drivers, and others who've historically lacked access to education and opportunity. Information networks have become a great leveler, and we should use them together to help lift people out of poverty and give them a freedom from want.[7]

The more recent successful examples of mobile phone applications, like M-Pesa—as of November 2009, half of Kenya's 16 million mobile phone users are also M-Pesa subscribers—have worked because they've been distinctly different from African Internet development projects in the past. First, there's been a sustainable and low-cost element to them, primarily through the deployment of profit-based models. Second, they are using a technology that is already widely deployed and doesn't require new levels of training and education. Third, there's an obvious and distinct advantage to using the new system, whereas it's not always immediately obvious why a produce vendor should have a webpage. The United States has influenced the creation of some of these tools, and it should continue to do so as part of its Twenty-first-Century Statecraft effort as an obvious way to use the power of the state to engage in political, economic, and public health development.

However, as the Internet becomes increasingly intertwined with international politics, it seems increasingly likely that the United States will come face-to-face with authoritarian regimes like Iran. The United States may stand for a "single Internet," but that vision simply does not exist today. There remains an Iranian Internet, where a great deal of information is blocked and filtered. In this case, it is naive to believe that the United States can engage in meaningful statecraft against countries that are not natural allies online. While YouTube videos from Obama are a nice Twenty-first-Century Statecraft gesture, they do little to change the Islamic Republic's behavior in any meaningful way.

Indeed, adversarial regimes have shown that they are fully capable of controlling and co-opting the Internet for their own purposes and are willing to intimidate, beat, and jail citizens who do not conform to the rules. Unless the United States is willing to violate the terms of traditional sovereignty, there isn't much incentive for Iran or other similar regimes to change their behavior. This is why Iran can act with such impunity on the world stage by blocking text messaging and filtering the Internet on a whim.

One way that has been proposed to overcome this impasse is through a very ancient piece of statecraft: the peace treaty. In fact, nine days after Secretary Clinton's speech in early 2010, the head of the International Telecommunications Union, Hamadoun Touré, called for an international treaty in which countries would agree not to engage in a first cyberstrike against one other. As the Agence France-Presse reported from the World Economic Forum in Davos, Switzerland:

> "A cyber war would be worse than a tsunami—a catastrophe," the UN official said, highlighting examples such as attacks on Estonia last year.
>
> He proposed an international accord, adding: "The framework would look like a peace treaty before a war."
>
> Countries should guarantee to protect their citizens and their right to access to information, promise not to harbour cyber terrorists and "should commit themselves not to attack another."[8]

It seems almost quaint to think that the world could sit down together and hammer out a peace treaty guaranteeing that countries would never attack each other online. Even if such a treaty were to be signed, the nature of a cyber-attack makes it quite likely that such attacks could continue while maintaining plausible deniability—any cyberattacker worth his salt knows how to cover up his digital tracks. Moreover, no nation could sign a cyberwar treaty with rogue non-state actors, like individual pro-regime hackers from places like Iran, China, or Russia.

When Iran or other authoritarian countries take steps to abuse their control of the Internet's infrastructure in their own countries, what can the United States (or any other country, for that matter) do about it? In other words, what happens when the twenty-first century actually meets statecraft with political adversaries? The answer, unfortunately, is very little. In the case of Iran, the United States can try to exert political pressure through sanctions and by forbidding American companies from doing business with Iran, and yet that has done little to change the Islamic Republic's behavior online. This is indeed a hard truth of the Internet, that local contexts matter a great deal more than most people realize. By the same token, the Iranian Internet is essentially created in the vision of the Islamic Republic, and despite the best efforts of dissidents to poke holes in its defenses, the fact remains that for all the dissident blogs, tweets, and YouTube videos, the regime remains in power. This is not to say that the Islamic Republic's vision will necessarily win out over American idealism—just that from elsewhere, the Internet looks and acts a lot differently than it does from here.

NOTES

INTRODUCTION

1. Veljo Haamer, in discussion with the author, June 14, 2009. Note: English is Haamer's third language, and he is prone to using words like "incomings," when in fact he means the English word "income."

2. "Iranian Satellite TV Channel Finds Inventive Ways to Broadcast Despite Censorship," June 23, 2009, Fox News, http://www.foxnews.com; Mir-Hossein Mousavi, Facebook status update, June 13, 2009, http://www.facebook.com/mousavi?v=feed&story_fbid=92079384588, translated by Lotfan.org, http://blog.lotfan.org/2009/06/chronology-of-june-13-2009.html.

3. Somayeh Tohidlou, FriendFeed message, June 13, 2009, http://friendfeed.com/smto/d47bd200.

4. Parastoo Dokouhaki, Twitter message, June 20, 2009, http://twitter.com/parastoo/status/2250977960; Babak Mehrabani, Twitter message, June 2009, http://twitter.com/BabakMehrabani/status/2250307699.

5. Daniel Terdiman, "Twitterverse Working to Confuse Iranian Censors," CNET News.com, June 16, 2009, http://news.cnet.com/8301-17939_109-10265462-2.html.

6. Web Ecology Project, "The Iranian Election on Twitter: The First Eighteen Days," June 26, 2009, http://webecologyproject.org/WEP-twitterFINAL.pdf.

7. "Iran Lifts Text Message Restrictions," July 1, 2009, United Press International, http://www.upi.com/Emerging_Threats/2009/07/01/Iran-lifts-text-message-restrictions/UPI-28421246472813/.

8. Mike Musgrove, "Twitter Is a Player in Iran's Drama," *Washington Post*, June 17, 2009, http://www.washingtonpost.com; Biz Stone, "Down Time Rescheduled," June 15, 2009, http://blog.twitter.com/2009/06/down-time-rescheduled.html.

9. Anonymous technical source in Tehran, in Skype text conversation with the author, June 22, 2009; Evgeny Morozov, "More on the Unintended Consequences of DDOS Attacks on Pro-Ahmadinejad Web-Sites," *Foreign Policy Net Effect,* June 18, 2009, http://neteffect.foreignpolicy.com/posts/2009/06/18/more_on_the_unintended_consequences_of_ddos_attacks_on_pro_ahmadinejad_web_sites.

10. Cyrus Farivar, "Twitter Confusion in Iran," June 25, 2009, *The World,* http://www.theworld.org.

11. Hamid Tehrani, "Iran's Revolutionary Guards Take on the Internet," January 8, 2009, http://blogs.law.harvard.edu/idblog/2009/01/08/irans-revolutionary-guards-take-on-the-internet/.

12. Gerdab.ir, July 1, 2009, http://74.125.53.132/translate_c?hl=en&ie=UTF-8&sl=fa&tl=
en&u=http://gerdab.ir/fa/pages/%3Fcid%3D422&prev=_t&rurl=translate.google
.com&usg=ALkJrhiV4th-amGhAh8bpUzPL3kqdHxkWA.

13. Austin Heap, "How to Set Up a Proxy for Iran Citizens," June 15, 2009, http://blog
.austinheap.com/how-to-setup-a-proxy-for-iran-citizens/.

14. Cyrus Farivar, "Geeks Around the Globe Rally to Help Iranians Online," July 8, 2009,
Tehran Bureau, http://www.pbs.org/wgbh/pages/frontline/tehranbureau/2009/07/
geeks-around-the-globe-rally-to-help-iranians-online.html.

15. Richard Esguerra, "Help Protestors in Iran: Run a Tor Bridge or Tor Relay," June 29,
2009, Deeplinks Blog, http://www.eff.org/deeplinks/2009/06/help-protesters-
iran-run-tor-relays-bridges; "Configuring a Tor Relay," June 30, 2009, http://www
.torproject.org.

16. Andrew Lewman, in discussion with the author, July 2, 2009. However, one downside
of Tor is that it slows down web traffic. Plus, it can be a little difficult to configure for
those who don't have a decent amount of technical knowledge, despite the fact that
the Tor website has many of its instructions and other webpages available in Persian
and other languages.

17. Vito Pilieci, "Ottawan Helps Iranians Bypass Firewall," June 20, 2009, Ottawa Citizen,
http://www.ottawacitizen.com; Ronald Deibert and Rafal Rohozinski, "Ottawa Needs
a Strategy for Cyberwar," June 30, 2009, National Post, http://network.nationalpost
.com/np/blogs/fullcomment/archive/2009/06/30/ottawa-needs-a-strategy-for-
cyberwar.aspx.

18. https://bbc-tweeter.net/a.php?token=a7d1b0d253174da9a221915c5b0acfb2.

19. Nart Villeneuve, in correspondence with the author, July 7, 2009.

20. "Iran's Biggest Ever Bourse Deal," Tehran Times, September 28, 2009, http://www
.tehrantimes.com.

21. Nicholas Negroponte, "Being Digital—A Book Preview," February 1, 1995, http://web
.media.mit.edu/~nicholas/Wired/WIRED3–02.html.

22. Captain George O. Squier, "The Influence of Submarine Cables upon Military and
Naval Supremacy," National Geographic 12, no. 1 (January 1901), 2.

23. Tom Standage, The Victorian Internet (New York: Walker, 1998), 104

24. John Markoff, "A Free and Simple Computer Link," New York Times, December 8, 1993,
http://www.nytimes.com.

25. Nasrin Alavi, We Are Iran: The Persian Weblogs (New York: Softskull, 2005), 361.

26. Golnaz Esfandiari, "Iran: Unemployment Becoming a 'National Threat,'" Radio Free
Europe, March 12, 2004, http://www.rferl.org; "The Populist's Problem," The Economist,
May 5, 2009, http://www.economist.com; "Iran Economy Facing 'Perfect Storm,'" BBC
News, October 24, 2008, http://news.bbc.co.uk.

27. "Population," Statistical Centre of Iran, http://www.sci.org.ir/portal/faces/public/sci_
en/sci_en.Glance/sci_en.pop; Scott MacLeod, "Our Veils, Ourselves," Time, July 27,
1998, http://www.time.com.

28. "Broadband Use Tracks Household Income—US Broadband Penetration at 40.9%—
November 2003 Bandwidth Report," November 26, 2003, http://www.websiteopti-
mization.com; Ken Belson with Matt Richtel, "America's Broadband Dream Is Alive in
Korea," New York Times, May 5, 2003, http://www.nytimes.com.

CHAPTER 1   SOUTH KOREA

Epigraph: Nagy Hanna, Sandor Boyson, and Shakuntala Gunaratne, "The East Asian Miracle and Information Technology: Strategic Management of Technological Learning," World Bank, 1996, http://www-wds.worldbank.org.

1. Bruce Wallace, "Gamer Is Royalty in S. Korea," *Los Angeles Times*, March 21, 2007, A-1, http://www.latimes.com. From here on, I will refer to South Korea simply as "Korea."

2. Wallace, "Gamer Is Royalty in S. Korea"; Nielsen Media Research, "Top TV Ratings," http://www.nielsenmedia.com/nc/portal/site/Public/menuitem.43afce2fac27e890311 baoa347a062ao/?vgnextoid=9e4df9669fa14010VgnVCMl00000880a260aRCRD.

3. Korean names are written with the family name first, followed by the first name.

4. Martyn Williams, "US Video Game Sales Sink in June, Biggest Drop in 9 Years," July 16, 2009, IDG News Service, http://www.pcworld.com/printable/article/id,168592/printable.html.

5. "Breaking the Ice: South Korea Lifts Ban on Japanese Culture," December 7, 1998, http://web-japan.org/trends98/honbun/ntj981207.html.

6. Daniel Lee, in discussion with the author, April 25, 2007.

7. Kim Ki-tae, "Will Starcraft Survive Next 10 Years?" *Korea Times*, March 20, 2005, http://web.archive.org/web/20050404025107/http://times.hankooki.com/lpage/culture/200503/kt2005032018105511690.htm.

8. Wallace, "Gamer Is Royalty in S. Korea."

9. Byungho Park, "'PC-Bang' Brought a 'Big-Bang': The Unique Aspect of the Korean Internet Industry," *Interface* 3, no. 8 (November 2003), http://209.85.173.132/search?q=cache:p7waSZ6sFoUJ:bcis.pacificu.edu/journal/2003/08/park.php+http://bcis.pacificu.edu/journal/2003/08/park.php&hl=en&ct=clnk&cd=1&gl=us. In English, this is often transliterated as "PC Bang." I have added an extra "h" to the transliteration to make the sound of the Korean word a little clearer.

10. "NCsoft Earnings Release, Q1, 2008," 10, http://www.ncsoft.net.

11. Parvez Hassan, "Korea: Problems and Issues in a Rapidly Growing Economy" (Washington, D.C.: International Bank for Reconstruction and Development/The World Bank, 1976), 26–27.

12. Ibid., 3–4; "Background Note: South Korea," U.S. Department of State, October 2008, http://www.state.gov.

13. Andrea Matles Savada, *South Korea: A Country Study* (Darby, Pa.: Diane Publishing, 1997), 157; Dr. Li Choy Chong, "Lecture 9: The Development of Infrastructures in Asia," page 15, http://www.arc.unisg.ch/org/arc/web.nsf/f1d9ed904340e51ac1256a8d00504269/50ae698667e7dbcec1256d2c00506b8e/$FILE/Lecture%209%20-%20Infrastructural%20developments.pdf; Dan Roem, "As Technology Changes, Students Find New Ways to Beat the System," *Gainesville Times*, April 23, 2008, http://www.wtop.com/?nid=25&pid=0&sid=1392213&page=1.

14. Ministry of Information and Communication and National Internet Development Agency of Korea, "Survey on the Computer and Internet Usage (Executive Summary)" (Seoul: Ministry of Information and Communication and National Internet Development Agency of Korea, 2004), 3; "Broadband Growth and Policies in OECD Countries," OECD 2008, 28–29: http://www.oecd.org.

15. Moon Ihlwan, "E-Society: My World Is Cyworld," *BusinessWeek*, September 26, 2005, http://www.businessweek.com; Justin Ewers, "Cyworld: Bigger Than YouTube?" *US News and World Report*, November 9, 2006: http://www.usnews.com. However, when the Cyworld site expanded to the United States in early 2007, it didn't catch on nearly as well. By mid-2008, Cyworld US only had 500,000 users, and Hyun-Oh Yoo, the CEO of SK Communications, Cyworld's parent company, called its American results "very disappointing" in a personal interview with the author. Cyworld US later closed down in November 2008.

16. Brandon Griggs, "Obama Poised to Be First 'Wired' President," CNN, January 15, 2009, http://www.cnn.com; David Carr, "Electoral Triumph Built on a Web Revolution," *International Herald Tribune*, November 9, 2008, http://www.iht.com; Jose Antonio Vargas, "e-Hail to the Chief," *Washington Post*, December 31, 2008, http://www.washingtonpost.com.

17. Hwang Kuhn, "The 2002 Presidential Election and Media Politics," *Korea Journal* 43, no. 2 (summer 2003): 201–202.

18. Mary Joyce, "The Citizen Journalism Web Site 'OhmyNews' and the 2002 South Korean Presidential Election," Berkman Center Research Publication No. 2007–15, December 2007, http://cyber.law.harvard.edu/publications/2007/The_Citizen_Journalism_Web_Site_Oh_My_News_and_the_South_Korean_Presidential_Election; Daniel Jisuk Kang and Laurel Evelyn Dyson, "Internet Politics in South Korea: The Case of Rohsamo and Ohmynews," paper presented at the 18th Australiasian Conference on Information Systems, December 5–7, 2007, Toowoomba, Australia; Oh Yeon Ho, "The End of 20th Century Journalism," *OhmyNews*, June 1, 2004, http://english.ohmynews.com/articleview/article_view.asp?article_class=8&no=169396&rel_no=1.

19. Dorothy Perkins, *Japan Goes to War: A Chronology of Japanese Military Expansion from the Meiji Era to the Attack on Pearl Harbor, 1868–1941* (Darby, Pa.: Diane Publishing, 1997), 83.

20. Ibid.

21. Ramon H. Myers et al., *The Japanese Colonial Empire, 1895–1945* (Princeton, N.J.: Princeton Univ. Press, 1987), 50; Harry Wray, *Japan Examined: Perspectives on Modern Japanese History* (Honolulu: Univ. of Hawaii Press, 1983), 226; John Lie, *Zainichi (Koreans in Japan): Diasporic Nationalism and Postcolonial Identity* (Berkeley: Univ. of California Press, 2008), 4; Minorities at Risk Project, *Assessment for Koreans in Japan*, December 31, 2003, http://www.unhcr.org/refworld/docid/469f3aa11e.html (accessed November 10, 2009).

22. Yang Sung Chul, "Student Political Activism: The Case of the 1960 April Revolution in South Korea," *Youth and Society* 8, no. 3 (1977): 299–320.

23. Chalmers Johnson, *Blowback* (New York: Macmillian, 2004), 25.

24. Bela A. Balassa and Marcus Noland, *Japan in the World Economy* (Washington, D.C.: Peterson Institute, 1988), 158.

25. Chon Kilnam, in discussion with the author, June 26, 2008.

26. Edward Sagendorph Mason, *The Economic and Social Modernization of the Republic of Korea: And Others* (Cambridge, Mass.: Harvard University Asia Center, 1980), 95; Michael Breen, *The Koreans: Who They Are, What They Want, Where Their Future Lies* (New York: Macmillan, 1999), 134.

27. Youngil Lim, *Technology and Productivity: The Korean Way of Learning and Catching Up* (Cambridge, Mass.: MIT Press, 1999), 93, 97; Lee Byeong-cheon and Pyŏng-ch'ŏn Yi,

Developmental Dictatorship and the Park Chung Hee Era: The Shaping of Modernity in the Republic of Korea (Paramus, N.J.: Homa and Sekey Books, 2005), 90; Yong-lob Chung, *South Korea in the Fast Lane: Economic Development and Capital Formation* (New York: Oxford Univ. Press, 2007), 10.

28. Chon Kilnam, in discussion with the author, June 26, 2008.

29. Ibid.

30. Stanford Computer Science, "About the Computer Science Department," Stanford University, http://www.cs.stanford.edu; Computer Science Division, "EECS History," University of California, Berkeley, http://www.eecs.berkeley.edu.

31. Chon Kilnam, in discussion with the author, April 8, 2007.

32. Leonard Kleinrock, "The Birth of the Internet," August 27, 1996, http://www.lk.cs .ucla.edu/LK/Inet/birth.html.

33. Scott Griffin, "Vint Cerf," Internet Pioneers website, iBiblio, http://www.ibiblio .org/pioneers/cerf.html.

34. Leonard Kleinrock, "The Day the Infant Internet Uttered Its First Words," March 9, 2005, http://www.lk.cs.ucla.edu/first_words.html.

35. Chon Kilnam, in discussion with the author, June 26, 2008.

36. Linsu Kim, "Stages of Development of Industrial Technology in a Developing Country: A Model," *Research Policy* 9 (1980): 266; Pervez Hasan, *Korea: Problems and Issues in a Rapidly Growing Economy* (Baltimore: World Bank, by Johns Hopkins Univ. Press, 1976), 177.

37. Sudip Chaudhuri, "Government and Economic Development in South Korea, 1961–79," *Social Scientist* 24, nos. 11/12 (November–December 1996): 18–35; Cheng-Fen Chen and Graham Sewell, "Strategies for Technological Development in South Korea and Taiwan: The Case of Semiconductors," *Research Policy* 25 (1996): 2; John A. Matthews and Dong-Sung Cho, *Tiger Technology: The Creation of a Semiconductor Industry in East Asia* (Cambridge: Cambridge Univ. Press, 2000), 116; Soon Il Ahn, "A New Program in Cooperative Research Between Academia and Industry in Korea, Involving Centers of Excellence," *Technovation* 15, no. 4 (May 1995): 241–257. KIST was actually proposed by President Lyndon Johnson in 1965 as a way to bolster American support for South Korea and to encourage Korean support of the Americans in Vietnam. The U.S. government put up 30 percent of the initial expenses for KIST during its first four years. See Seong-Rae Park, *Science and Technology in Korean History*, trans. Seong-Rae Park (Freemont, Calif.: Jain, 2005), 280–281.

38. Park, *Science and Technology in Korean History*, 280–281; Chen and Sewell, "Strategies for Technological Development," 6.

39. Keith Oblitas and J. Raymond Peter, "Transferring Irrigation Management to Farmers in Andhra Pradesh, India," World Bank technical paper no. 449 (Washington, D.C.: World Bank Publications, 1999), 93.

40. Park, *Science and Technology in Korean History*, 282; Moo-Young Han, "Annotated Chronology of Korea's Science and Technology," http://www.duke.edu/~myhan/ kafo401.html; Youngil Lim, *Technology and Productivity: The Korean Way of Learning and Catching Up* (Cambridge, Mass.: MIT Press, 1999), 135; Oblitas and Peter, "Transferring Irrigation Management to Farmers in Andhra Pradesh, India," 93.

41. Matthews and Cho, *Tiger Technology*, 116.

42. Lee Yong-Teh, in discussion with the author, July 16, 2008.

43. Ibid.

44. Hasan, *Korea: Problems and Issues in a Rapidly Growing Economy,* 185.

45. Lee Yong-Teh, in discussion with the author, July 16, 2008; Petr L. Spurney Jr., "The World Bank and the Republic of Korea, 1962–1994," http://www-wds.worldbank.org/external/default/WDSContentServer/WDSP/IB/2005/05/17/000090341_20050517125638/Rendered/INDEX/312810KR0Productive1partnership0ıpublicı.txt; U.S. Congress, Office of Technology Assessment, *Competing Economies: American, Europe and the Pacific Rim* (Darby, Pa.: Diane Publishing, 1991), 317.

46. Penelope Francks et al., *Agriculture and Economic Development in East Asia: From Growth to Protectionism in Japan, Korea, and Taiwan* (New York: Routledge, 1999), 133.

47. Spurney, "The World Bank and the Republic of Korea, 1962–1994."

48. Globalis—South Korea, http://globalis.gvu.unu.edu/indicator_detail.cfm?IndicatorID=19&Country=KR.

49. "Office Computers," Historic Computers in Japan website, Information Processing Society of Japan, http://www.ipsj.or.jp/katsudou/museum/history/history_mf_e.html,.

50. Adrian Buzo, *The Making of Modern Korea* (New York: Routledge, 2002), 24, 32.

51. Jang-Sup Shin, *The Economics of the Latecomers: Catching-Up, Technology Transfer, and Institutions in Germany, Japan, and South Korea* (New York: Routledge, 1996), 134.

52. Youngil Lim, *Technology and Productivity: The Korean Way of Learning and Catching Up* (Cambridge, Mass.: MIT Press, 1999), 141.

53. ARPANET Geographic Map, 1980, http://personalpages.manchester.ac.uk/staff/m.dodge/cybergeography/atlas/arpanet4.gif.

54. Chon Kilnam, in discussion with the author, April 2005.

55. Ibid., April 7, 2009.

56. Robert Wade, *Governing the Market: Economic Theory and the Role of Government in East Asia Industrialization* (Princeton, N.J.: Princeton Univ. Press, 2004), 314; Mathews and Cho, *Tiger Technology,* 120.

57. Chon Kilnam, e-mail message to the author, August 28, 2008; Chon Kilnam, in discussion with the author, June 26, 2008.

58. Chon Kilnam, "Subject: Software Development Network(SDN)—Preliminary," September 30, 1981.

59. Kilnam Chon and Kee Wook Rim, "Computer System Development with Standardization," *North Holland Computers and Standards* 2 (1983): 177.

60. Chon Kilnam, in discussion with the author, April 7, 2009.

61. Oblitas and Peter, "Transferring Irrigation Management to Farmers in Andhra Pradesh, India," 93.

62. Park Hyunje, in discussion with the author, July 15, 2008.

63. Ibid.

64. Hur Jin Ho, in discussion with the author, April 9, 2007.

65. Lee Dongman, in discussion with the author, September 1, 2008.

66. Yoon Kim, in discussion with the author, September 1, 2008; Lee Dongman, in discussion with the author, September 1, 2008.

67. Chon Kilnam, in discussion with the author, April 2007.

68. Ibid.

69. Hur Jin Ho, in discussion with the author, April 9, 2007.

70. Ibid., June 23, 2008.

71. Park Hyunje, in discussion with the author, April 19, 2007.

72. Chon Kilnam, e-mail to Larry Landweber, December 21, 1984.

73. Park Hyunje, in discussion with the author, April 19, 2007.

74. Park Hyunje, e-mail to Torben, March 24, 1990.

75. Lee Dongman, in discussion with the author, September 1, 2008.

76. In cases where Koreans use English names, I have chosen to write them in a Western style, with the first name first.

77. Hur Jin Ho, in discussion with the author, April 9, 2007.

78. Ibid.

79. Ibid.

80. "Office Computers," Historic Computers in Japan website.

81. Jim Larson, e-mail message to the author, November 13, 2008.

82. Lucien Rhodes, "The Race for Bandwidth," *Wired* 4, no.1 (January 1996), http://www.wired.com/wired/archive/4.01/medin_pr.html.

83. Park Hyunje, in discussion with the author, April 19, 2007.

84. Kyounglim Yun et al., "The Growth of Broadband Internet Connections in South Korea: Contributing Factors" (Stanford, Calif.: Asia Pacific Research Center, Stanford University, 2002), 17.

85. Lee Yong-Teh, in discussion with the author, July 16, 2008.

86. Cheon JoUn, "A Comprehensive Plan for Building the Korea Information Infrastructure," *Proceedings of the 12th International Conference on Computer Communication on Information Highways: For a Smaller World and Better Living* (Amsterdam: IOS Press, 1996): 15–20.

87. Jyoti Choudrie and Heejin Lee, "Broadband Development in South Korea: Institutional and Cultural Factors," *European Journal of Information Systems* 13 (2004): 103–114.

88. Yang Seung-taik, "Digital Divide and Cyber Korea 21 Initiative" (New York: United Nations, 2002), 1.

89. Heekyung Hellen et al., "Broadband Penetration and Participatory Politics: South Korea Case," *Proceedings of the 37th Hawaii International Conference on System Sciences—2004,* Honolulu, Hawaii, 4; Choudrie and Lee, "Broadband Development in South Korea," 103–114; Kim Min-bai, "President Kim Sends First E-mail to Cabinet," *Chosun Ilbo,* February 2, 2000, http://english.chosun.com/w21data/html/news/200002/200002020536.html.

90. Jake Song, in discussion with the author, July 18, 2008.

91. Blizzard Entertainment Press Release, "StarCraft's 10-Year Anniversary: A Retrospective," March 31, 1998, http://web.archive.org/web/20080405155309/http://www.blizzard.com/us/press/10-years-starcraft.html; Kelly Olsen, "South Korea Gamers Get a Sneak Peek at StarCraft II," Associated Press, May 21, 2007, http://www.usatoday.com/tech/gaming/2007-05-21-starcraft2-peek_N.htm; Chon Kilnam, "A Brief History of the Internet in Korea."

92. Andrew Pollack, "Koreans Place Kia Motors under Bankruptcy Shield," *New York Times,* July 16, 1997, http://www.nytimes.com; Vikas Bajaj, "Stocks Soar in Seoul on Kia Takeover Plan," *New York Times,* October 23, 1997, http://www.nytimes.com; Timothy L. O'Brien and Andrew Pollack, "Korea Situation Deteriorates, Raising

Specter of a Default," *New York Times,* December 12, 1997, http://www.nytimes.com; Thomas Crampton, "Seoul, Bangkok and Jakarta Hit: 3 East Asia Nations Get 'Junk' Ratings," *International Herald Tribune,* December 23, 1997, http://www.nytimes.com/1997/12/23/news/23iht-junk.t.html?pagewanted=print.

93. J. C. Herz, "The Bandwidth Capital of the World," *Wired* 10, no. 8, http://www.wired.com; Florence Chee, "Essays on Korean Online Gaming /A Sense of Place: Media and Motivation in Korea by the Wang-Tta Effect /Order and Chaos in an Ethnography of Korean Online Game Communities," Simon Fraser University, Burnaby, British Columbia, 9.

94. Huhh Jun-sok, "Culture and Business of PC Bang in Korea," August 1, 2007, http://papers.ssrn.com/so13/papers.cfm?abstract_id=975171, 4; Byungho Park, "PC-Bang! How Internet Cafes Sparked the Amazing Growth of the Korean Information Technology Industry" (paper presented at the annual meeting of the International Communication Association, TBA, San Francisco, Calif., May 23, 2007); Moon Ihlwan, "The Champs in Online Games," *BusinessWeek,* July 23, 2001, http://www.businessweek.com.

95. Howard W. French, "2 Korean Leaders Speak of Making 'A Day in History,'" *New York Times,* June 14, 2000, http://www.nytimes.com.

96. Hong Seongtae, e-mail message to the author, November 23, 2008.

97. Heekyung et al., "Broadband Penetration and Participatory Politics: South Korea Case," 4; Sung-Tae Hong, "Rohsamo, New Political Movement in South Korea," http://blog.peoplepower21.org/English/7522.

98. Howard W. French, "Online Newspaper Shakes up Korean Politics," *New York Times,* March 6, 2003, http://www.nytimes.com.

99. Doc Searls, "Printwash, Cont'd," http://doc-weblogs.com/2003/05/19.

100. Jean Min, in discussion with the author, April 12, 2007; Jean Min, "Participatory Journalism in Action: The Cases of Ohmynews.com, Wikinews.org, and the Blogs," April 8, 2005, International Symposium on Online Journalism, University of Texas, Austin, http://online.journalism.utexas.edu/videos.php?year=2005.

101. "OhmyNews Leads Reform of Civic Journalism," *Newspapers and Technology,* January 2006, http://www.newsandtech.com.

102. "Case: Ohmynews," 2005, http://www.slideshare.net/marketingfacts/case-ohmynews.

103. Nicholas Lemann, "Amateur Hour," *New Yorker,* August 7, 2006, http://www.newyorker.com/archive/2006/08/07/060807fa_fact1?printable=tru.

104. Don Kirk, "Koreans Protest U.S. Military's Handling of a Fatal Accident," *New York Times,* August 4, 2002, http://www.nytimes.com; "Seoul Restaurants Bar US Diners," BBC News, November 28, 2002, http://news.bbc.co.uk.

105. Sergeant Russell C. Bassett, "Court-Martial Begins Today," Release #5–2002118, November 18, 2002, http://8tharmy.korea.army.mil/PAO/releases/trial%20day%201.doc; James Brooke, "First of 2 G.I.'s on Trial in Deaths of 2 Korean Girls Is Acquitted," *New York Times,* November 21, 2002, http://www.nytimes.com; Ronda Hauben, "Online Grassroots Journalism and Participatory Democracy in South Korea," 8, http://www.columbia.edu/~rh120/other/netizens_draft.pdf.

106. Yonghoi Song, "Internet News Media and Issue Development: A Case Study on the Roles of Independent Online News Services as Agenda-Builders for Anti-US Protests in South Korea," *New Media Society* (2007): 80–81.

107. Je HunHo, in discussion with the author, April 24, 2007; Marco Evers, "South Korea Turns PC Gaming into a Spectator Sport," trans. Chistropher Sultan, *Der Spiegel,* June 2, 2006, http://www.spiegel.de/international/spiegel/0,1518,druck-399476,00.html.

108. Daniel Lee, in discussion with the author, April 25, 2007.

109. KESPA, "e-Sports in Korea," May 2008; "White Style," http://www.whitestyle.com/.

110. Elizabeth Woyke, "Asia's Smart Metropolis," *Forbes,* September 21, 2009, http://www.forbes.com.

111. Pamela Licalzi O'Connell, "Korea's High-Tech Utopia, Where Everything Is Observed," *New York Times,* October 5, 2005, http://www.nytimes.com.

112. Carl Seaholm, in discussion with the author, June 22, 2007.

113. Sean Campbell, "Economic Development: Metropolis from Scratch," *Next American City,* April 2005, http://americancity.org/magazine/article/economic-development-metropolis-from-scratch-campbell/.

114. Zach Mortice, "New Songdo City Looks Back at the New World," *AIA Architect,* July 25, 2008, http://www.aia.org/aiarchitect/thisweek08/0725/0725d_songdo.cfm; Campbell, "Economic Development."

115. Campbell, "Economic Development."

116. Paulo Alcazaren, "Korea Builds City of the Future—Today," *Philippine Star,* November 21, 2009, http://www.philstar.com/Article.aspx?articleId=525273&publicationSubCategoryId=85.

117. Campbell, "Economic Development."

118. Anthony Faiola, "When Escape Seems Just a Mouse-Click Away," *Washington Post,* May 27, 2006, http://www.washingtonpost.com.

119. Victoria Kim, "Video Game Addicts Concern South Korean Government," Associated Press, October 6, 2005, http://www.usatoday.com/tech/gaming/2005–10–06-korean-game-addicts_x.htm; Phillippe Naughton, "Korean Drops Dead after 50-Hour Marathon," *Times* (London), August 10, 2005, http://www.timesonline.co.uk; Choe Sang-Hun, "Hooked on the Virtual World: A Reality in South Korea," *International Herald Tribune,* June 12, 2006, http://www.iht.com; Vanessa Hua, "Video Game Players Score Big Money in South Korea," *San Francisco Chronicle,* December 18, 2006, http://www.sfgate.com/cgi-bin/article.cgi?f=/c/a/2006/12/18/GAMERS.TMP; Choe Sang-Hun, "Hooked on the Virtual World"; Faiola, "When Escape Seems Just a Mouse-Click Away."

120. Jerald J. Block, M.D., "Issues for DSM-V: Internet Addiction," *American Journal of Psychiatry* 165 (March 2008): 306–307, http://ajp.psychiatryonline.org/cgi/content/full/165/3/306#R1653BABHGAGF.

121. Kim Hyesoo, in discussion with the author, April 18, 2007.

122. Elizabeth Woyke, "Grappling with Internet Addiction," *Forbes,* April 3, 2009, http://www.forbes.com; Martin Fackler, "In Korea, a Boot Camp Cure for Web Obsession," *New York Times,* November 18, 2007, http://www.nytimes.com.

123. "Study: Kids at Risk for Internet Addiction," United Press International, October 25, 2008, http://www.upi.com/Health_News/2008/10/25/Study_Kids_at_risk_for_Internet_addiction/UPI-49741224958883/; Kim, "Video Game Addicts Concern South Korean Government."

124. Kim, "Video Game Addicts Concern South Korean Government."

125. "South Korean Children Face Gaming Curfew," BBC, April 13, 2010, http://news.bbc.co.uk; "S Korea Child 'Starves as Parents Raise Virtual Baby,'" BBC, March 5, 2010, http://news.bbc.co.uk.

126. Block, "Issues for DSM-V: Internet Addiction."

127. Jonathan Krim, "Subway Fracas Escalates into Test of the Internet's Power to Shame," *Washington Post,* July 7, 2005, http://www.washingtonpost.com; "낚시로 방송 낚았다," dkb news, July 6, 2005, http://www.dkbnews.com/main.php?mn=news&mode=read&nidx=7735.

128. Daniel Solove, "Of Privacy and Poop: Norm Enforcement via the Blogosphere," Balkinization website, June 30, 2005, http://balkin.blogspot.com/2005/06/of-privacy-and-poop-norm-enforcement.html.

129. Moon Ihlwan, "Korea's U.S. Beef Brouhaha," *BusinessWeek,* June 9, 2008, http://www.businessweek.com.

130. Michael Hurt, "Mad about Mad Cow Disease in Korea," Scribblings of the Metropolitician website, May 5, 2008, http://metropolitician.blogs.com/scribblings_of_the_metrop/2008/05/mad-about-mad-c.html; "Mad Cow Thesis Twisted Out of All Proportion," *Chosun Ilbo,* May 9, 2008, http://english.chosun.com/w21data/html/news/200805/200805090017.html.

131. Jon Herskovitz and Rhee So-eui, "South Korean Internet Catches 'Mad Cow Madness,'" Reuters, June 13, 2008, http://in.reuters.com/article/internetNews/idINSE030506420080613?sp=true.

132. "MBC Needs to Acknowledge Its Mistakes," *Chosun Ilbo,* May 21, 2008, http://english.chosun.com/w21data/html/news/200805/200805210025.html.

133. Choe Sang-Hun, "Protests in Seoul More about Nationalism Than Beef," *International Herald Tribune,* June 11, 2008, http://www.iht.com/articles/2008/06/11/asia/seoul.php.

134. Choe Sang-Hun, "An Anger in Korea over More Than Beef," *New York Times,* June 12, 2008, http://www.nytimes.com; "S. Korea Delays U.S. Beef Imports," CNN, June 2, 2008, http://www.cnn.com; Tom Walsh, "Web Hysteria a Danger to Korean Deal," *Detroit Free Press,* June 16, 2008, http://www.juneauempire.com/stories/061608/opi_291413665.shtml; "Going Overboard with Mad Cow Scare," *Chosun Ilbo,* May 2, 2008, http://english.chosun.com/w21data/html/news/200805/200805020019.html.

135. Ronda Hauben, "Korean Netizens Continue Demonstrations," *Telepolis,* June 7, 2008, http://www.heise.de/tp/r4/artikel/28/28085/1.html; Robert J. Koehler, "Gyopo ID'd for Uploading Fake Video of Protest Crackdown," The Marmot's Hole website, May 28, 2008, http://www.rjkoehler.com/2008/05/28/gyopo-idd-for-uploading-fake-video-of-protest-crackdown/; "South Korea Jails Man for Internet Lies during Beef Protests," Agence France-Presse, October 23, 2008, http://timesofindia.indiatimes.com/World/SKorea_jails_man_for_internet_lies/articleshow/3631392.cms.

136. Choe Sang-Hun, "Korean Leader Considers Ways to Rework Government," *New York Times,* June 11, 2008, http://www.nytimes.com.

137. "SKorea's Lee Highlights Internet Benefits, Dangers," Agence France-Presse, June 17, 2008, http://afp.google.com/article/ALeqM5gmGVnclG3B49_9aj9FhdSqEf2ZSg.

138. Michael Fitzpatrick, "South Korea Braced for Web Clampdown," *The Guardian,* August 5, 2008, http://www.guardian.co.uk.

139. Michael Fitzpatrick, "South Korea Wants to Gag the Noisy Internet Rabble," *The Guardian,* October 8, 2008, http://www.guardian.co.uk.

140. "Clean Slate: An Interdisciplinary Research Program," http://cleanslate.stanford.edu/.

141. Chon Kilnam, in discussion with the author, August 3, 2008.

CHAPTER 2   SENEGAL

Epigraph: U.S. Agency for International Development (USAID), "Senegal Telecommunications Sector Assessment," Washington, D.C., January 1998.

1. Ibrahima Yock, in discussion with the author, January 2007.

2. Chiffres clés, "Tarifs des liaisons ADSL de la SONATEL, 1 octobre 2008," Observatoire sur les Systèmes d'Information, les Réseaux et les Inforoutes au Sénégal, http://www.osiris.sn/article462.html.

3. "Présentation des résultats de l'enquête nationale sur les technologies de l'information et de la communication au Sénégal /ENTICS 2009," ARTP, April 1 2010, http://www.artp-senegal.org/page_inter2.php?idmenu=5&id=10104&sid=10104&art=319&label=News&page=Enqu%EAte+nationale+sur+les+TIC+2009.

4. Chiffres clés, "Internet," Observatoire sur les Systèmes d'Information, les Réseaux et les Inforoutes au Sénégal, http://www.osiris.sn/article27.html; CIA, "The World Factbook— Senegal," https://www.cia.gov.

5. Olivier Sagna, "Lutte contre la fracture numérique: Passer du discours aux actes," Editorial, *Batik,* April 1, 2010, http://www.osiris.sn/article87.html.

6. OSIRIS stands for Observatoire sur les Systèmes d'Information, les Réseaux et les Inforoutes au Sénégal (Observatory on Information Systems, Networks and Information Highways in Senegal).

7. Amadou Top, in discussion with the author, March 4, 2007.

8. Ibid.

9. Ibid.

10. J. D. Fage et al, *The Cambridge History of Africa* (Cambridge: Cambridge Univ. Press, 1977), 457; Bethwell Ogot, ed., *General History of Africa,* vol. 5, *Africa from the Sixteenth Century to the Eighteenth Century* (Oxford, U.K.: Currey, 1999), 6.

11. Ibid., 139.

12. Olivier Sagna, "Information and Communication Technologies and Social Development in Senegal: An Overview," trans. Paul Keller, 7, http://www.unrisd.org/80256B3C005BCCF9/(httpPublications)/BA28329C8E73447D80256B5E0037AB04?OpenDocument; Cheikh Tidiane Gadio, "Institutional Reform of Telecommunications in Senegal, Mali, and Ghana: The Interplay of Structural Adjustment and International Policy Diffusion" (Ph.D. diss., Ohio State University, 1995), 180.

13. Sagna, "Information and Communication Technologies," 42.

14. Gadio, "Institutional Reform of Telecommunications in Senegal, Mali, and Ghana," 183.

15. Jean-Paul Azam et al, "Telecommunications Sector Reforms in Senegal," World Bank Development Research Group Regulation and Competition Policy, September 2002, 6–7.

16. U.S. Agency for International Development (USAID), "Senegal Telecommunications Sector Assessment," Washington, D.C., January 1998. 15.

17. Sagna, "Information and Communication Technologies," 42; Peter Benjamin, "African Experience with Telecenters," *e-OnTheInternet,* November/December 2000, http://www.isoc.org/oti/articles/1100/benjamin.html.

18. Banque Centrale des Etats de l'Afrique de l'Ouest, "History of the CFA Franc," http://www.bceao.int/internet/bcweb.nsf/pages/umuse1.

19. Barry James, "In Africa, Both Money and Paris' Role Shrink," *International Herald Tribune,* January 24, 1994, http://www.iht.com; Kenneth B. Noble, "French Devaluation of African Currency Brings Wide Unrest," *New York Times,* February 23, 1994, http://www.nytimes.com.

20. Azam et al., "Telecommunications Sector Reforms in Senegal," 24; Sonatel, "Historique," trans. Cyrus Farivar, http://sonatel.sn/xamxam,1,0,1e.

21. Amadou Top, "Enfin ! ! !," trans. Cyrus Farivar and Ben Akoh, *Batik* 54 (January 2004), http://osiris.sn/article892.html#Enfin%20!!!.

22. "Senegal Awards Third Mobile License to Sudatel," Reuters, September 8, 2008, http://asia.news.yahoo.com/070907/3/37ko2.html; Aly Diouf, "Sudatel négligerait le Sénégal," *Wal Fadjri,* May 6, 2008, http://www.osiris.sn/article3604.html.

23. Mouhamet Diop, e-mail message to Steven Huter, October 30, 1996, http://nsrc .org/db/lookup/report.php?id=890202362168:497430065&fromISO=SN.

24. Ibid.; Cyrus Farivar, "The Powerful Force: An Examination of the Internet in Senegal," April 28, 2004, http://cyrusfarivar.com/docs/thesis_final.html; "Michel Mavros et Oumou Sy," *Africultures,* July 2002, http://www.africultures.com/index.asp?menu= affiche_murmure&no_murmure=177.

25. Blaise Rodriguez, in discussion with the author, November 1, 2002.

26. USAID Leland Initiative, "Leland Initiative: Africa GII Gateway Project / Project Description & Frequently Asked Questions," Africa Global Information Infrastructure Project, http://www.usaid.gov/leland/project.htm.

27. Ibid.; USAID, "Senegal Telecommunications Sector Assessment," January 1998, 26.

28. Amadou Top, in discussion with the author, March 10, 2007.

29. "Randy Bush and the NSRC," Network Startup Resource Center, http://ran.psg.com/ ~randy/nsrc.html; Larry Landweber, in discussion with the author, March 28, 2007.

30. USAID, "Senegal Telecommunications Sector Assessment," FOIA Executive Summary, January 1998, 1.

31. U.S. Embassy Dakar cable to U.S. Department of Commerce, June 11, 1997, 3.

32. USAID, "Senegal Telecommunications Sector Assessment," FOIA Executive Summary, January 1998, 1.

33. "Senegal: Parliamentary Chamber: Assemblée nationale," Inter-Parliamentary Union, http://www.ipu.org/parline-e/reports/arc/2277_98.htm; Moussa Paye, "New Information Technologies and the Democratic Process," trans. Victoria Bawtree, UNRISD, May 2002, 13, http://www.unrisd.org/80256B3C005BCCF9/(httpPublications)/C8BC7D4551 751920C1256D73005972B4?OpenDocument.

34. Amadou Top, in discussion with the author, March 10, 2007.

35. Paye, "New Information Technologies and the Democratic Process," 13.

36. Amadou Top, "OSIRIS un lieu d'échange et de discussion," trans. Cyrus Farivar, *Batik* 1 (August 1999), OSIRIS, http://www.osiris.sn/article302.html#OSIRIS%20un%20lieu% 20d%27%C3%A9change%20et%20de%20discussion.

37. "Network Information Center," NIC Senegal, http://www.nic.sn/index.php?option= com_content&task=view&id=20&Itemid=35; Farivar, "The Powerful Force."

38. "Press Conference by UNICEF Goodwill Ambassador," UNICEF, October 23, 2000, http://www.un.org/News/briefings/docs/2000/20001023.ndourprc.doc.html.

39. Janine Firpo, in discussion with the author, March 23, 2007.

40. Jennifer Schenker, "Building Bridges," *Time,* February 8, 2001, http://www.time.com; Nicholas Thompson, "Logging On: Can Africa's Most Popular Musician Get His Country Online?" *Boston Globe,* October 14, 2002, http://www.newamerica.net/publications/articles/2002/logging_on_can_africas_most_popular_musician_get_his_country_online; Lisa Goldman, "The Joko Project," http://209.85.173.132/search?q=cache:u3a7x0v7EYgJ:www.iterations.com/protected/dwnload_files/joko_execsum.doc+joko+club&hl=en&ct=clnk&cd=2&gl=us&client=firefox-a.

41. Lisa Carney and Janine Firpo, "Joko Pilot Results in Senegal," *TechKnowLodgia,* issue 17, http://www.techknowlogia.org/TKL_active_pages2/CurrentArticles/main.asp?IssueNumber=17&FileType=HTML&ArticleID=417.

42. Thompson, "Logging On: Can Africa's Most Popular Musician Get His Country Online?"

43. Jeremiah Hall, "Dot-com Exiles Turn Up at New Charities," *Christian Science Monitor,* June 23, 2003, http://www.globalpolicy.org/socecon/crisis/2003/0623dot.htm.

44. Janine Firpo, in discussion with the author, March 23, 2007.

45. Loi No. 2001–15 du 27 décembre 2001 /Portant /Code de Télécommunications, page 1, Government of Senegal, Dakar, Senegal (trans. Cyrus Farivar).

46. Ben Akoh, "The Case for 'Open Access' Communications Infrastructure in Africa: The SAT-3/WASC Cable /Senegal Case Study," Association for Progressive Communications, 9, http://www.apc.org/en/pubs/research/openaccess/africa/case-open-access-communications-infrastructure-afr.

47. Ibid.

48. Aliou Kane Ndiaye, "Malick Guèye DG de l'Agence de regulation des telecommunications (ART): 'L'ART est totalement indépendante,'" *Nouvel Horizon,* February 18, 2005, http://www.osiris.sn/article1609.html; "Etat des lieux," Agence de Régulation des Télécommunications et des Postes, http://www.artp-senegal.org/page_inter2.php?idmenu=3&id=10083&sid=10085&art=44&label=&page=Secteur+des+T%E91%E9coms; Akoh, "The Case for 'Open Access' Communications Infrastructure in Africa," 28.

49. Ousmane Mbaye, in discussion with the author, January 18, 2007.

50. "Sacked Senegalese Regulator Questioned by DIC over Embezzlement Charges," *Balancing Act,* issue 267, http://www.balancingact-africa.com/news/back/balancing-act_267.html#top.

51. "Senegal: La barrière du monopole," Panos Institute West Africa, Dakar, Senegal, August 1, 2002, http://www.panos-ao.org/ipao/spip.php?article2983.

52. USAID letter to Amadou Top, August 1, 2001, 1.

53. Ibid., 4.

54. "Toutes les Etapes," Caravane Multimédia, http://web.archive.org/web/20011122120155/http://www.caravane-multimedia.sn/etapes.asp; Marian Zeitlin, *The EcoYoff News letter,* issue 18 (September 2001): 3.

55. Mamadou Gaye, in discussion with the author, January 12, 2003.

56. Ibid.

57. "Senegal: La barrière du monopole," The PANOS Institute West Africa, August 1, 2002, http://www.panos-ao.org/ipao/spip.php?article2983.

58. USAID letter to Amadou Top, August 1, 2001, 5–6.

59. "Peace Corps Director to Help Launch Digital Freedom Initiative," *Peace Corps Online,* March 5, 2003, http://peacecorpsonline.org/messages/messages/2629/2012330.html.

60. Andrew S. Natsios, "Digital Freedom Initiative Remarks," March 4, 2003, http://www.usaid.gov/press/speeches/2003/ty030304.html.

61. DFI Senegal Design, USAID, Washington, D.C., August 14, 2003, 3.

62. Ibid., 7.

63. DFI Final Report, USAID, Washington, D.C., August 14, 2003, 2, 7.

64. Matt Berg, in discussion with the author, January 16, 2007.

65. Anonymous, in discussion with the author, January 3, 2007.

66. United Nations, "United Nations Millennium Declaration," September 8, 2000, http://www.un.org/millennium/declaration/ares552e.htm.

67. Abdoulaye Wade, "Allocution," trans. Cyrus Farivar, December 11, 2003, http://www.itu.int/wsis/geneva/coverage/statements/senegal/sn-fr.html.

68. Ibid.

69. Anick Jesdanun, "U.N. Summit Ends with Net Expansion Plans," Associated Press, December 12, 2003, http://www.informationweek.com/news/showArticle.jhtml?articleID=16700456.

70. Ambassador David A. Gross, "The Success of the UN World Summit on the Information Society and What It Bodes for the Future," December 16, 2003, http://fpc.state.gov/27379.htm.

71. Cheikh Alioune Jaw, "Amadou Top: 'La démonopolisation dans les telecommunications sera une excellente chose,'" trans. Cyrus Farivar, *Nouvel Horizon,* February 6, 2004, http://www.osiris.sn/article899.html; Sonatel, "Télécommunications: fin du monopole de la SONATEL," trans. Cyrus Farivar, ART Press Release, July 19, 2004.

72. Thierno Ousmane Sy, "Thierno Ousmane Sy au Grand Jury de RFM," trans. Cyrus Farivar, September 30, 2007, http://www.infotechsn.com/index.php/2008042839/Multimedia/Podcast-audio/Thierno-Ousmane-Sy-au-Grand-Jury-de-RFM.html; "Le Conseil special du chef de l'Etat chargé des Ntic, Thierno Ousmane Sy raconte," trans. Cyrus Farivar, *Nettali,* September 13, 2007, http://www.nettali.net/spip.php?article 4735.

73. Amadou Top, "Doing Business . . . ," trans. Cyrus Farivar, *Batik,* issue 98 (October 1, 2007), http://www.osiris.sn/article3086.html#Doing%20Business. . . .

74. Sonatel Annual Report 2007, 15, www.sonatel.sn.

75. Global Digital Solidarity Fund Annual Review 2007–2008, 35, www.dsf-fsn.org.

76. Ibid., 31.

77. Jean Pouly, in discussion with the author, December 1, 2008.

78. Ababacar Diop, in discussion with the author, December 1, 2008.

79. Robert Walter Lawler and Masoud Yazdani, *Artificial Intelligence and Education* (New York: Ablex Publishing Corporation, 1987), 7, http://books.google.com/books?hl=en&lr=&id=iCbbXxnpMoUC&oi=fnd&pg=PP11&dq=senegal+negroponte+Dray&ots=PIqpKULR0i&sig=eWdVFK4CY5U1W6ysTlFeW5IJrkY#PPA7,M1; Deborah Gage, "Solar-Powered Wi-Fi a Gift to Senegal," *San Francisco Chronicle,* August 16, 2008, http://www.sfgate.com/cgi-bin/article.cgi?f=/c/a/2008/08/16/BUA712BH10.DTL.

80. Donald Z. Osborn, e-mail to A12n-forum, "Microsoft Goes Wolof," December 5, 2004, http://lists.kabissa.org/lists/archives/public/a12n-forum/msg00254.html.

81. Association for Progressive Communications, "SAT3: What Happens When National Monopolies End, and What Does This Mean for Policy-Makers and Regulators?" May 2007, 4, 5, www.apc.org.

82. Orange Senegal, "ADSL 512," http://www.orange.sn/djoumtouwayyatu,33,8,ay,do9dc4 214b2930b; Chiffres clés, "Internet"; "ICT at a Glance /Estonia," World Bank, November 2007, 1, www.worldbank.org.

83. Jean Pouly, in discussion with the author, December 1, 2008.

84. Sharon LaFranière, "Cellphones Catapult Rural Africa to 21st Century," *New York Times,* August 25, 2005, http://www.nytimes.com; "Senegalese Mobile Base Tops Five Million Users; Tigo Fights for License," C114.net, November 13, 2008, http://www.cn-c114 .net/576/a361475.html.

85. Amadou Top, "Libéralisation ne rime pas avec libération," trans. Cyrus Farivar, *Batik,* issue 110 (September 2008), http://osiris.sn/article3939.html#Lib%C3%A9ralisation %20ne%20rime%20pas%20avec%20lib%C3%A9ration. . . .

CHAPTER 3   ESTONIA

Epigraph: Jaanus Friis, "The U.S. President gets a Skype Phone," November 28, 2006, http://share.skype.com/sites/en/2006/11/the_us_president_gets_a_skype.html.

1. Veljo Haamer, in discussion with the author, February 7, 2007.

2. Soumitra Dutta, *The Global Information Technology Report 2006–2007*: Connecting to the Networked Economy (New York: Palgrave Macmillan, 2007), 174; "President Bush Welcomes President Ilves of Estonia to the White House," Office of the Press Secretary, June 25, 2007, http://georgewbush-whitehouse.archives.gov/news/releases/2007/06/ 20070625.html.

3. Mike Collier, "Ilves Warns against 'Wishful Thinking,'" *Baltic Times,* July 26, 2007, http://www.baltictimes.com.

4. Mart Laar, "The Estonian Economic Miracle," Heritage Foundation, August 7, 2007, http://www.heritage.org/Research/WorldwideFreedom/bg2060.cfm.

5. Ibid.

6. Andreas Thomann, "Credit Suisse Discovers the 'Baltic Tigers,'" July 13, 2006, http:// emagazine.credit-suisse.com/app/article/index.cfm?aoid=159202&fuseaction= OpenArticle&lang=en.

7. Adalbert Knöbl, Andres Sutt, and Basil Zavoico, "The Estonian Currency Board: Its Introduction and Role in the Early Success of Estonia's Transition to a Market Economy" (May 2002), IMF Working Paper No. 02/96, available at SSRN: http://ssrn .com/abstract=879651.

8. Peter Parkes, "Skype's Share of International Calling Minutes Jumps 50%," Skype blog, January 19, 2010, http://share.skype.com/sites/en/2010/01/skypes_share_of_ international.html.

9. Jason Begay, "Escaping to Bryant Park, But Staying Connected to the Web," *New York Times,* July 3, 2002, http://www.nytimes.com.

10. Jacob Farkas, in discussion with the author, August 8, 2007.

11. Ibid.

12. Veljo Haamer, in discussion with the author, January 25, 2008.

13. Esther B. Fien, "Baltic Citizens Link Hands to Demand Independence," *New York Times,* August 24, 1989, http://www.nytimes.com.

14. Veljo Haamer, in discussion with the author, March 15, 2005.

15. Ivo Kivinurk, in discussion with the author, July 12, 2007.

16. Heiki Kübbar, in discussion with the author, February 6, 2008.

17. Veljo Haamer, in discussion with the author, March 15, 2005.

18. Toivo Raun, *Estonia and the Estonians* (Stanford, Calif.: Hoover Press, 2001), 3, 16.

19. Ibid., 38.

20. Ibid., 71.

21. Ibid., 92.

22. Doudou Diene, "Racism, Racial Discrimination, Xenophobia, and Related Forms of Intolerance, Follow-up to and Implementation of the Durban Declaration and Programme of Action," United Nations Human Rights Council, March 17, 2008, 6, www.un.org.

23. Jaak Kangilaski et al., *The White Book: Losses Inflicted on the Estonian Nations by Occupation Regimes (1940–1991)* (Tallinn: Republic of Estonia, 2005), 11.

24. Diene, "Racism, Racial Discrimination, Xenophobia and Related Forms of Intolerance," 6.

25. Vladimir Raudsepp, *Estonia '78* (Tallinn, Estonia: Perioodika, 1978), 99.

26. Jacques Rupnik, Vello Pettai, and Jan Zielonka, *The Road to the European Union* (Manchester: Manchester Univ. Press, 2003), 260.

27. Kangilaski et al., *The White Book,* 187.

28. Tönu Parming and Elmar Jarvesoo, *A Case Study of a Soviet Republic: The Estonian SSR* (Boulder, Colo.: Westview, 1978), title page.

29. "Estonia, The Transition to a Market," World Bank, 1993, xv, http://www-wds .worldbank.org/external/default/WDSContentServer/WDSP/IB/1993/03/01/000009265_ 3970128103057/Rendered/INDEX/multiopage.txt.

30. Rupnik, Pettai, and Zielonka, *The Road to the European Union,* 79.

31. Mart Laar, in discussion with the author, July 19, 2007.

32. "Estonia, Implementing the EU Accession Agenda," World Bank, 1999, ix, www.world-bank.org.

33. "Mart Laar's Biography," Cato Institute, http://www.cato.org/special/friedman/laar/index .html; Sheryl Gay Stolberg, "In Estonia, Bush Finds a Tax Actually to His Liking," *New York Times,* November 30, 2006, http://www.nytimes.com.

34. Laar, "The Estonian Economic Miracle."

35. Meelis Kitsing, "The Impact of Economic Openness on Internet Diffusion in Slovenia and Estonia" (Master's thesis, Tufts University, Medford, Mass., 2004), 22; "Estonia: Living Standards during the Transition: A Poverty Assessment," World Bank, 1996, http://go.worldbank.org/7LoTW814Yo.

36. Allar Viivik, "Eesti sai internetti enne Valget Maja," *Õhtuleht newspaper,* April 6, 2002, www.ohtuleht.ee.

37. Ibid.

38. Mart Laar, in discussion with the author, July 19, 2007.

39. Hanno Haamer, in discussion with the author, July 20, 2007.

40. Rainer Nõlvak, in discussion with the author, February 21, 2008.

41. "Eesti Telekom Subsidiary Purchases IT-Provider MicroLink," press release, May 17, 2005, http://www.telekom.ee/index.php?lk=1004&uudis=202.

42. Peeter Parvelo, e-mail message to the author, June 21, 2008.

43. Ibid.

44. Jaak Aaviksoo, in discussion with the author, September 19, 2007.

45. Kalle Uibo, "President Meri and Businessmen Go to the US to Prepare Project Tigerleap," trans. Maria Visnapuu, *Äripäev*, November 6, 1995, www.aripaev.ee.

46. The European Union Phare program is designed to assist the applicant nations from Central and Eastern Europe to ascend to the European Union. See Toomas Mattson, "Tiger Takes Only a Bite of 35.5 Million for the Next Year," trans. Maria Visnapuu, *Postimees*, September 24, 1996, www.postimees.ee; "Tiger Leap 1997–2007," Tiger Leap Foundation, 41, http://www.tiigrihype.ee/?op=body&id=97.

47. Inno Tähismaa, "The Number of Permanent Internet Connection Users Triples Every Year," trans. Maria Visnapuu, *Äripäev*, October 17, 1996, www.aripaev.ee.

48. "Tiger Leap Foundation Established, First Public Internet Station Opened," trans. Maria Visnapuu, *Postimees*, February 22, 1997, www.postimees.ee.

49. Andrew Meier, "Estonia's Tiger Leap to Technology," UNDP, http://web.archive .org/web/20080121084529/http://www.sdnp.undp.org/it4dev/stories/estonia.html.

50. Shel Israel, "Linnar Viik and Internet Road Sign," October 23, 2006, http://www .flickr.com/photos/shelisrael/278570253/in/set-72157594341308492/; Kristjan Otsmann, "Internet Reaches the People," trans. Maria Visnapuu, *Eesti Päevaleht*, February 12, 1997.

51. "Internet ja NATO," *Postimees*, February 24, 1997, http://www.postimees.ee/leht/ 97/02/22/uudis.htm#kolmas.

52. Peeter Parvelo, e-mail message to the author, June 21, 2008.

53. Hanno Haamer, in discussion with the author, February 18, 2008.

54. Ibid.

55. Veljo Haamer, in discussion with the author, March 15, 2005.

56. "Tigerleap II Is a Good Idea," trans. Maria Visnapu, *Äripäev*, March 26, 1998, www.aripaev.ee.

57. Mart Laar, in discussion with the author, July 19, 2007.

58. Linnar Viik, in discussion with the author, March 15, 2005.

59. "e-Estonia: The Internet Penetrates Every Aspect of Private and Public Life in Estonia, From Banking Services to Local Elections," International Special Reports, http://web.archive.org/web/20070914041926/http://www.internationalspecial reports.com/europe/01/estonia/banking_finance+/; "Online Biz Is Booming in Estonia," Associated Press, April 21, 2003, http://www.wired.com/print/culture/lifestyle/news/ 2003/04/58561.

60. Linnar Viik, in discussion with the author, March 15, 2005; Tarvi Martens, in discussion with the author, July 11, 2007.

61. "Identity Documents Act (1999)," Estonian Parliament, http://web.archive.org/web/2007 1023172337/http://www.legislationline.org/legislation.php?tid=11&lid=2278&less=false; Andres Aarma, "Estonian ID-card," Estonian Certification Center, 9, www.sk.ee.

62. Tarvi Martens, e-mail message to the author, February 26, 2008; Andres Aarma, in discussion with the author, July 11, 2007; "The Success Story of the Electronic Tax Board," idBlog, March 7, 2007, http://www.id.ee/blog_en/?p=9; "Nearly All Estonia's Taxpayers File Returns via Net," Agence France-Presse, April 9, 2010, http://www .google.com/hostednews/afp/article/ALeqM5j-iRZfRrgs8NVxp4b5MfDTHLU6Ew.

63. John Borland, "Online Voting Clicks in Estonia," *Wired News,* March 2, 2007, http://www.wired.com/print/politics/security/news/2007/03/72846.

64. Ibid.

65. Veljo Haamer, in discussion with the author, April 17, 2008.

66. Glenn Fleishman, "The Web without Wires, Wherever," *New York Times,* February 22, 2001, http://www.nytimes.com.

67. Veljo Haamer, "WiFi in Estonia: The Current Situation," *Baltic IT&T Review,* issue 34 (2004), 42.

68. Veljo Haamer and Anto Veldre, "WiFi in Estonia: The Current Situation," *Baltic IT&T Review,* issue 34 (2004), 13.

69. "Ameerika Rahukorpus lõpetab tegevuse Eestis," *Pressiteade,* June 17, 2002, http://estonia.usembassy.gov/rahukorpus.php.

70. "Skype Beta Launched," August 29, 2003, http://about.skype.com/2003/08/skype_beta_launched.html; Ben Charny, "Newsmaker: Why VoIP Is Music to Kazaa's Ear," CNET News, September 11, 2003, http://www.news.com/2008-1082–5074558.html.

71. Taavet Hinrikus, in discussion with the author, July 23, 2007.

72. Jaanus Friis, "Origin of the Name/Word 'Skype,'" Skype.com, November 1, 2005, http://share.skype.com/sites/en/2005/11/origin_of_the_nameword_skype.html.

73. Charny, "Newsmaker: Why VoIP Is Music to Kazaa's Ear."

74. Steven Woehrel, "Estonia: Current Issues and U.S. Policy," CRS Report for Congress, July 11, 2007, http://fas.org/sgp/crs/row/RS22692.pdf.

75. Cyrus Farivar, "Soviet Union Communist Party Forever! #2, Tallinn," July 16, 2007, http://www.flickr.com/photos/cfarivar/851525083/in/set-72157600893723380/.

76. "Dvigatel," NTI, March 2009, http://nti.org/e_research/profiles/Estonia/Nuclear/4736_4887.html; André Veskimeister, in discussion with the author, July 17, 2007.

77. "Gunnar Kobin Steps Down as CEO of Kesko Food after Only 6 Months," *Äripäev,* August 13, 2003, http://www.aripaev.ee/2444/free/summary.html; Toomas Hõbemägi, "Ulemiste City Borrows EEK 315 mln from a Bank Syndicate," *Baltic Business News,* September 5, 2006, http://www.balticbusinessnews.com.

78. "Ülemiste City Launched a Link 'Job Offers,'" Ülemiste City press release, February 2008, http://www.ulemistecity.ee/eng/news/?newsID=2080.

79. "Estonia Seals off Soviet Memorial," BBC News, April 26, 2007, http://news.bbc.co.uk.

80. Steven Lee Meyers, "Estonia: Excavation Near Soviet Monument," *New York Times,* April 27, 2007, http://www.iht.com/articles/2007/04/27/news/estonia.php; "Second Night of Rioting Hits Tallinn," *Baltic Times,* April 28, 2007, http://www. baltictimes.com.

81. Mark Lander and John Markoff, "Digital Fears Emerge after Data Siege in Estonia," *New York Times,* May 29, 2007, http://www.nytimes.com.

82. Joshua Davis, "Hackers Take Down the Most Wired Country in Europe," *Wired,* August 21, 2007, http://www.wired.com/print/politics/security/magazine/15–09/ff_estonia.

83. Urmas Vahe, "On the Front Lines of an Invisible War," *Baltic IT&T Review,* November 3, 2007, http://www.ebaltics.com/00704599; Mirjam Mäekivi, "Viik: küberrünnak suveräänse riigi vastu oli maailmas esmakordne," *Postimees,* May 17, 2007, http://www.postimees.ee/170507/esileht/siseuudised/261227.php.

84. Mirjam Mäekivi, "Viik: küberrünnak suveräänse riigi vastu oli maailmas esmako-rdne," *Postimees,* May 17, 2007, http://www.postimees.ee/170507/esileht/siseuudi-sed/261227.php; Heli Tiirmaa-Klaar, in discussion with the author, May 14, 2008; Lauri

Almann, "Kaitseministeeriumi kantsler Lauri Almanni ettekanne Washingtonis mainekal sisejulgeolekukonverentsil GOVSEC 2008 Eesti kogemusest küberjulgeoleku tagamisel," Estonian Ministry of Defense, May 8, 2008, http://www.mod.gov.ee/?op=body&id=490.

85. Hillar Aarelaid, in discussion with the author, July 17, 2007; Silver Meikar, "Küberrünnak Eesti vastu leiab kajastust kõikjal (TÄIENDATUD)," Meikar.ee, May 21, 2007, http://www.meikar.ee/blog/?p=305; General Johannes Kert, in discussion with the author, May 5, 2008.

86. Any computer on the Internet has a unique number, known as an IP address. This can be used as a way to trace where online traffic is originating from. That said, this number can also be easily faked.

87. Hillar Aarelaid, in discussion with the author, July 17, 2007.

88. Margus Kurm, in discussion with the author, July 19, 2007.

89. "Tulemused—Teksid," Estonian Ministry of Justice, Tallinn, http://www.legaltext.ee/et/andmebaas/tekst.asp?loc=text&dok=X30068K7&keel=en&pg=1&ptyyp=RT&tyyp=X&query=karistusseadustik.

90. Margus Kurm, in discussion with the author, July 19, 2007.

91. Margus Kurm, e-mail message to the author, January 29, 2008.

92. Toomas Hendrik Ilves, "Address by President Toomas Hendrik Ilves to the 62nd Session of the UN General Assembly," September 25, 2007, http://www.un.estemb.org/statements_articles/aid-546.

93. Kenneth Geers, in discussion with the author, May 5, 2008.

94. Ibid.

95. John William Parker, "Literacy in the Baltics," University of Washington, June 2002, http://depts.washington.edu/baltic/papers/literacy.htm; Jaak Peetre and Jaan Penjam, eds., "Semigroups and Automata. Selecta Uno Kaljulaid (1941–1999)," vol. 0 (2006), http://www.booksonline.iospress.nl/Content/View.aspx?piid=5337.

96. Kauksi Ülle et al., "Ethno-Futurism as a Mode of Thinking for an Alternative Future," trans. Sven-Erik Soosaar, May 5, 1994, http://www.suri.ee/etnofutu/ef!eng.html.

97. Tanel Raun, "The Development of Estonian Literacy in the 18th and 19th Centuries," *Journal of Baltic Studies* (June 1, 1979): 124.

98. Ibid, 125.

99. Michael Castells and Pekka Himanen, *The Information Society and the Welfare State: The Finnish Model* (Oxford, U.K.: Oxford Univ. Press, 2004), 132.

100. Laar, "The Estonian Economic Miracle."

101. Castells and Himanen, *The Information Society and the Welfare State*, 55.

102. Linnar Viik, in discussion with the author, June 11, 2008.

103. Andres Langemets, "The Estonian Tiger Leap into the 21st Century, Appendix 1," Estonian Information Society, http://www.esis.ee/ist2000/background/tiigrihype/app1.html.

104. Ott Ummelas and Rishaad Salamat, "Estonia Ready to Join Euro in 11 Months, Moody's Says," *Bloomberg,* January 21, 2010, http://www.bloomberg.com.

105. Christian Buck, "Fear, Web Posting in Estonia," *Wired News,* August 27, 1999, http://www.wired.com/techbiz/media/news/1999/08/21428; "Estonia Shames Errant Fathers in Cyberspace," Agence France-Presse, January 29, 2010, http://www.google.com/hostednews/afp/article/ALeqM5g0_yVOs6GRy33rKn9HbhGvFPUIsA.

106. "What Is eGA?" eGovernance Academy, http://www.ega.ee/en/eGAabout.

107. Veljo Haamer, in discussion with the author, January 25, 2008; Colin Woodward, "Estonia, Where Being Wired Is a Human Right," *Christian Science Monitor,* July 1, 2003, http://www.csmonitor.com; Meier, "Estonia's Tiger Leap to Technology."

CHAPTER 4   IRAN

Epigraph: William Gibson, "A Talk Given at the Directors Guild of America's Digital Day," Los Angeles, May 17, 2003, http://www.williamgibsonbooks.com/blog/2003_05_01_archive.asp.

1. "Youth Could Sway the Vote in Iran Election," December 6, 2009, Euronews, http://www.euronews.net/2009/06/12/youth-could-sway-the-vote-in-iran-election/.

2. Omid Memarian, in discussion with the author, August 27, 2007, and December 7, 2007.

3. John Kelly and Bruce Etling, "Mapping Iran's Online Public: Politics and Culture in the Persian Blogosphere," Berkman Center for Internet and Society, April 6, 2008, 2; Chuck Olsen, "Hoder on Iranian Blogs," March 12, 2005, http://blogumentary.type-pad.com/chuck/2005/03/hoder_on_irania.html; Ben Macintyre, "Mullahs versus the Bloggers," *The Times,* December 2005, http://www.timesonline.co.uk/tol/comment/columnists/ben_macintyre/article782133.ece; Raymond G. Gordon Jr., ed., *Ethnologue: Languages of the World,* 15th ed. (Dallas, Tex.: SIL International, 2005), online version: http://www.ethnologue.com/show_ country.asp?name=IR; David Sifry, "The State of the Live Web, April 2007," April 5, 2007, http://www.sifry.com/alerts/archives/000493.html.

4. Hamid Tehrani, "Iran's Revolutionary Guards Take on the Internet," Internet and Democracy Blog, January 8, 2009, http://blogs.law.harvard.edu/idblog/2009/01/08/irans-revolutionary-guards-take-on-the-internet/; Kelly and Etling, "Mapping Iran's Online Public," 6–7.

5. These figures are calculated based on data provided by the CIA Factbook, and taking the number of Internet users listed divided by the population.

6. "Iran Launches Huge Fibre-Optic Network," Associated Press, July 12, 2005, http://www.theage.com.au/news/breaking/iran-launches-huge-fibreoptic-network/2005/07/12/1120934207594.html.

7. "Internet Under Surveillance 2004," Reporters Sans Frontières, 2005, http://www.rsf.org/article.php3?id_article=10748.

8. Ibid.; "Internet Use Restored by Burmese Junta," Associated Press, October 15, 2007, http://www.thestar.com/News/article/266818.

9. Tim Wu and Jack Goldsmith, *Who Controls the Internet?* (Oxford, U.K.: Oxford Univ. Press, 2006), 91; "China's Middle Class," *The Economist,* January 17, 2002, http://www.economist.com; Amelia Newcomb, "China Goes to College—in a Big Way," *Christian Science Monitor,* July 29, 2005, http://www.csmonitor.com.

10. Clive Thompson, "Google's China Problem (and China's Google Problem)," *New York Times Magazine,* April 23, 2006, http://www.nytimes.com.

11. The OpenNet Initiative involves researchers at the Citizen Lab at the Munk Centre for International Studies, University of Toronto, the Berkman Center for Internet and Society at Harvard Law School, the Advanced Network Research Group at the Cambridge Security Programme, the University of Cambridge, and the Oxford Internet

Institute at Oxford University. ONI, "Internet Filtering in China in 2004–2005: A Country Study," April 13, 2005.

12. "Indo-European Telegraph Department," *Encyclopaedia Iranica,* http://www.iranica .com/newsite/articles/v13f1/v13f1010.html; Shahram Sharif, in discussion with the author, August 29, 2007; Mr. Behi, "Will You Ever Ask Me What I Like?" August 17, 2006, http://mrbehi.blogs.com/i/2006/08/will_you_ever_a.html.

13. Annabelle Sreberny-Mohammadi and Ali Mohammadi, *Small Media, Big Revolution: Communication, Culture, and the Iranian Revolution* (Minneapolis: Univ. of Minnesota Press, 1994), 119–120; Yassaman Taghi Beigi, "Tehran Dishes Out Satellite Crackdown," *Inter Press Service,* January 5, 2002, http://atimes.com/c-asia/DA05Ag02.html; "Iran Prohibits Satellite Dishes to Bar U.S. TV," *New York Times,* December 27, 1994, http://www.nytimes.com; Bob Garfield, "The News from Iran," *On the Media,* August 25, 2006, http://www.onthemedia.org/transcripts/2006/08/25/01.

14. Borzou Daragahi, "Iran: Surf's Up, in Tehran," *Babylon and Beyond,* October 25, 2007, http://latimesblogs.latimes.com/babylonbeyond/2007/10/surfs-up-in-teh.html; Hamid Tehrani, "Iran: YouTube and Facebook Are Not Filtered Anymore," *Global Voices Online,* February 7, 2009, http://globalvoicesonline.org/2009/02/07/iran-you-tube-and-face-book-are-not-filtered-anymore/.

15. "Handbook for Bloggers and Cyber-Dissidents," http://www.rsf.org/article.php3?id_ article=15050.

16. Geneive Abdo, "Online Ayatollah: Isolated Iranian Dissident Speaks Out on Web," *International Herald Tribune,* July 29, 2000, http://www.iht.com/articles/2000/07/29/ iran.2.t_8.php; "Montazeri.com vs. Montazery.com: Ayatollahs wage war on Internet," Agence France-Presse, December 15, 2000, http://web.payk.net/mailingLists/iran-news/html/2000/msg00477.html.

17. "Sentence Reduced, While Two More Arrested and 400 Internet Cafés Closed," *IFEX Communiqué* 10, no. 20 (May 21, 2001), http://www.ifex.org/20en/content/view/full/ 28726; "Iran Bars Web for Under-18s," Reuters, June 25, 2001, http://www.zdnet .com.au/news/soa/Iran-bars-Web-for-under-18s/0,139023165,120234327,00.htm; "Iran Annual Report 2002," *Press Freedom throughout the World,* April 24, 2002, http://www.rsf.org/article.php3?id_article=1438; "Internet Related Offenses Should Be Dealt with Separately," Islamic Republic News Agency, September 15, 2002, http://www.netnative.com/news/02/sep/1052.html.

18. "False Freedom: Online Censorship in the Middle East and North Africa," Human Rights Watch, November 2007, http://hrw.org/reports/2005/mena1105/5.htm.

19. Robert Tait, "Iran Bans Fast Internet to Cut West's Influence," *The Guardian,* October 18, 2006, http://www.guardian.co.uk.

20. Hamid Tehrani, "Bloggers on the Iran Government's Efforts to Enforce Registration of Blogs and Web Sites," January 10, 2007, http://www.globalvoicesonline.org/2007/ 01/10/bloggers-on-the-iran-governments-efforts-to-enforce-registration-of-blogs-and-web-sites/; Omid Memarian, "RIGHTS-IRAN: Bloggers Rebel at New Censorship," Inter Press Service News Agency, January 10, 2007, http://www.ipsnews.net/news .asp?idnews=36123.

21. "Iran," OpenNet Initiative, June 16, 2009, http://opennet.net/research/profiles/iran.

22. Siavash Shahshahani, e-mail message to the author, April 28 2010.

23. Carroll Bogert, "Chat Rooms and Chadors," *Newsweek,* August 21, 1995, 36.

24. Ibid.; Declan McCullagh, "Banning Iran," *HotWired,* August 28, 1996, http://www
    .eff.org/Misc/Publications/Declan_McCullagh/hw.banning.iran.082896.txt.

25. Muhammad Khatami, *Hope and Challenge: The Iranian President Speaks* (Binghampton,
    N.Y.: Institute of Global Cultural Studies, Binghampton University, 1997), 81–82.

26. Christiane Amanpour, "Transcript of Interview with Iranian President Mohammad
    Khatami," CNN, January 7, 1998, http://www.cnn.com/WORLD/9801/07/iran/interview
    .html.

27. Roschanack Shaery-Eisenlohr, "Hamyaran: The Iranian NGO Resource Center: An
    Interview with Baquer Namazi, Director of Hamyaran." *GSC Quarterly* 10 (fall 2003).

28. John F. Burns, "As Iran's Reformer Speaks, Anti-Reformers Sit and Scowl," *New York
    Times,* September 30 1999, http://www.nytimes.com.

29. "Iran—Constitution," http://www.servat.unibe.ch/law/icl/ir00000_.html.

30. Shahram Sokooti, "Voice of Iran? Political Pawn? Iran's Reformist Press," *World Press
    Review,* January 22, 2002, http://www.worldpress.org/mideast/0122iran.htm; Afshin
    Molavi, *The Soul of Iran* (New York: Norton, 2005), 130; Tara Mahtafar, "Mr.
    Mohajerani Goes to Washington," *Tehran Bureau,* October 20, 2009, http://www
    .pbs.org/wgbh/pages/frontline/tehranbureau/2009/10/mr-mohajerani-goes-to-
    washington.html.

31. "World: Middle East /Rafsanjani's Daughter Cleared," BBC, December 7, 1998,
    http://news.bbc.co.uk; "'Zan' Ordered Closed for Two Weeks," Agence France-Presse,
    January 23, 1999, http://web.payk.net/mailingLists/iran-news/html/1999/msg00271.
    html; "Attacks on the Press 1999: Iran," Committee to Protect Journalists, March 20,
    2000, http://www.cpj.org/attacks99/mideast99/Iran.html.

32. "Population," Statistical Centre of Iran, http://www.sci.org.ir/portal/faces/public/sci_
    en/sci_en.Glance/sci_en.pop.

33. DNI-News Digest—19 Jan 2000—Special issue, January 19, 2000, http://d-n-i.abdolian
    .com/news/dnd/2000/01/%5B00–01–19%5DDNI-NEWS%20Digest%20-%2019%20Jan%
    202000%20-%20Special%20issue.htm.

34. Masoud Safiri, in discussion with the author, November 12, 2007; Molavi, *The Soul of
    Iran,* 125.

35. The full list of these newspapers is: *Asr-e-Azadegan, Fat'h, Aftab-e-Emrooz, Arya,
    Gozaresh-e-Ruz, Bamdad-e-No, Payam-e-Azadi, Azad, Payam-e-Hajar, Aban, Arzesh, Iran-e-
    Farda, Sobh-e-Emrooz,* and *Akhbar Eqtesad;* "Press Crackdown Intensifies, Fourteen
    Newspapers Closed, Two Journalists Imprisoned," *CPJ/IFEX,* April 25, 2000, http://
    canada.ifex.org/es/content/view/full/10218.

36. Payman Arabshahi, "Iran's Telecom and Internet Sector: A Comprehensive Survey,"
    June 15, 1999, http://www.science-arts.org/internet/node5.html.

37. Hossein Derakhshan, in discussion with the author, September 12, 2007.

38. Salman Jariri, in discussion with the author, September 23, 2007.

39. Salman Jariri, "Salman's Weblog," trans. Rebekah Kouy-Ghadosh, September 10, 2001,
    http://209.85.173.132/search?q=cache:yv5DqZXqxugJ:www.globalpersian.com/archive/
    010901.html+http://www.globalpersian.com/archive/010901.html&cd=1&hl=en&ct=
    clnk&gl=us&client=safari.

40. Hossein Derakhshan, in discussion with the author, September 12, 2007.

41. Hossein Derakhshan, "Editor: Myself," October 2001, http://i.hoder.com/archives/
    2001/10/011007_005490.shtml.

42. Masoud Safiri, in discussion with the author, November 12, 2007.

43. Sanam Dolatshahi, in discussion with the author, October 19, 2007.

44. Sanam Dolatshahi, "Khorshid Khanoom," trans. Sanam Dolatshahi, November 9, 2001, http://www.khorshidkhanoom.com/archives/001282.php.

45. "False Freedom: Online Censorship in the Middle East and North Africa," Human Rights Watch, November 2007, http://hrw.org/reports/2005/mena1105/5.htm#_Toc119125722.

46. Omid Memarian, in discussion with the author, August 23, 2007.

47. Parvin Alizadeh, "Iran's Quandry: Economic Reforms and the 'Structural Trap,'" *Brown Journal of World Affairs* 9, no. 2 (winter/spring 2003): 270.

48. "Iran: Country Reports on Human Rights Practices," U.S. Department of State, February 25, 2005, http://www.state.gov/g/drl/rls/hrrpt/2004/41721.htm.

49. "Iran Election 'an Alarm Bell,'" BBC, March 3, 2003, http://news.bbc.co.uk.

50. Jim Muir, "Q&A: Iranian Election Row," BBC, February 10, 2004, http://news.bbc.co.uk.

51. Sina Motalebi declined to be interviewed for this book, as he continues to fear retribution against himself and/or his family still in Iran. All information about him and his story is gathered from other sources.

52. Omid Memarian, in discussion with the author, August 27, 2007.

53. Ibid.

54. Ibid.

55. Omid Memarian, April 25, 2003, trans. Omid Memarian, http://omemarian.blogspot.com/index.html#93234458.

56. Omid Memarian, May 5, 2003, trans. Omid Memarian, http://omemarian.blogspot.com/index.html#93798374.

57. Omid Memarian, in discussion with the author, August 27, 2007.

58. Ali Reza Eshraghi, in discussion with the author, June 1, 2009.

59. Omid Memarian, in discussion with the author, December 2, 2007.

60. Mohammad Khatami, "President of the Islamic Republic of Iran before the World Summit on the Information Society," December 10, 2003, http://www.itu.int/wsis/geneva/coverage/statements/iran/ir.html.

61. Aaron Scullion, in discussion with the author, November 20, 2007.

62. Hossein Derakhshan, "Who Said Technology Can't Help Democracy?" Hoder.com, December 11, 2003, http://hoder.com/weblog/archives/009147.shtml.

63. "World Summit on the Information Society," WSIS, December 11, 2003, http://www.itu.int/wsis/geneva/coverage/archive.asp?lang=en&c_type=pc|11.

64. Omid Memarian, in discussion with the author, December 1, 2007.

65. Aaron Scullion, "Iran's ICT Minister Confronted," *Daily Summit*, December 11, 2003, http://www.dailysummit.net/english/archives/2003/12/11/irans_ict_minister_confronted.asp.

66. Ali Akbar Dareini, "Iran's Judiciary Chief Orders Ban on Use of Torture by Interrogators," Associated Press, April 28, 2004, www.ap.org.

67. Hossein Derakhshan, "CIA Runs 'Spider's Web' in Iran, Radical Paper Claims," Hoder.com, September 24, 2004, http://hoder.com/weblog/archives/012304.shtml.

68. "Iran: Four Journalists Sentenced to Prison, Floggings," Human Rights Watch, February 10, 2009, http://www.hrw.org/en/news/2009/02/10/iran-four-journalists-sentenced-prison-floggings?print.

69. Omid Memarian, in discussion with the author, September 3, 2007.

70. Touran Memarian, in discussion with the author, trans. Rebekah Kouy-Ghadosh, November 5, 2007.

71. "Iran Cracks Down on Blog Protests," BBC, October 13, 2004, http://news.bbc.co.uk.

72. Shahram Rafizadeh, in discussion with the author, trans. Rebekah Kouy-Ghadosh, November 29, 2007.

73. "Crackdown on Internet Journalists in Iran," Reuters, October 13, 2004, http://www.signonsandiego.com/news/world/20041013–0626-media-iran-arrests.html.

74. Omid Memarian, in discussion with the author, September 3, 2007.

75. Ibid.

76. Omid Memarian, in discussion with the author, December 2, 2007.

77. "Iranian Vice President Resigns," Associated Press, October 12, 2004, http://www.usatoday.com/news/world/2004–10–12-iran-vp_x.htm.

78. Parsithan, "Imprisoned Bloggers Were Tortured, Says ex-VP," Persian Students in the United Kingdom, December 28, 2004, http://www.persianstudents.org/archives/001243.html; Nazila Fathi, "An Iranian Cleric Turns Blogger for Reform," New York Times, January 16, 2005, http://www.nytimes.com.

79. Fathi, "An Iranian Cleric Turns Blogger for Reform."

80. "Iranian Cleric Blogs for Free Expression," Reuters, February 16, 2005, http://yaleglobal.yale.edu/display.article?id=5288.

81. "Human Rights Watch Honors Iranian Journalist and Blogger," Human Rights Watch, October 26, 2005, http://www.hrw.org/en/news/2005/10/26/human-rights-watch-honors-iranian-journalist-and-blogger.

82. Christopher de Bellaigue, The Struggle for Iran (New York: New York Review Books, 2007), 71.

83. Omid Memarian, in discussion with the author, December 2, 2007.

84. Nazila Fathi, "The Iranian Exile's Eye," New York Times, January 10, 2010, http://www.nytimes.com.

85. Afshin Molavi, in discussion with the author, December 4, 2007.

86. Juan Cole, "Hitchens the Hacker; And, Hitchens the Orientalist; And, 'We don't Want Your Stinking War!,'" May 3, 2006, http://www.juancole.com/2006/05/hitchens-hacker-and-hitchens.html; Arash Norouzi, "'Wiped Off the Map': Rumor of the Century," http://www.mohammadmossadegh.com/news/rumor-of-the-century/; "Iranian Leader: Holocaust a 'Myth,'" CNN, December 14, 2005, http://www.cnn.com; Mike Shuster, "Iran Enriches Uranium, Plans New Expansion," NPR, April 11, 2006, http://www.npr.org/templates/story/story.php?storyId=5336802.

87. Mahmoud Ahmadinejad, "Autobiography," President Ahmadinejad's Blog, August 8, 2006, http://www.ahmadinejad.ir/en/autobiography/.

88. Mahmoud Ahmadinejad, "Merry Christmas to Everyone," President Ahmadinejad's Blog, December 21, 2006, http://www.ahmadinejad.ir/en/merry-christmas-to-everyone/.

89. Mahmoud Ahmadinejad, "A Guideline for Islamic Governance," President Ahmadinejad's Blog, December 1, 2007, http://www.ahmadinejad.ir/en/a-guideline-for-islamic-governance/.

90. Cyrus Farivar, "The President's Infrequent Blog," *The World,* December 5, 2007, http://theworld.org/?q=node/14491; Geoff Siskind, "This Week's Show (Dec.13/07). Links & Music," *Search Engine,* December 13, 2007, http://www.cbc.ca/search engine/blog/2007/12/this_weeks_show_dec1307_links_1.html.

91. Robert Tait, "Iran's Clerics Caught up in Blogging Craze," *The Guardian,* October 11, 2006, http://www.guardian.co.uk; Hamid Tehrani, "Iranian Muslim Bloggers," History News Network, November 26, 2007, http://hnn.us/articles/44774.html. Hamid Tehrani is a pseudonym.

92. "Iran: Four Journalists Sentenced to Prison, Floggings."

93. Omid Memarian, in discussion with the author, December 2, 2007.

94. Wu and Goldsmith, *Who Controls the Internet?,* 180.

95. Omid Memarian, in discussion with the author, December 17, 2007.

CONCLUSION

1. Hillary Clinton, "Remarks on Internet Freedom," January 21, 2010, http://www.state.gov/secretary/rm/2010/01/135519.htm.

2. Alec Ross, "U.S. Diplomacy in the Age of Facebook and Twitter: An Address on 21st Century Statecraft," December 18, 2009, http://www.state.gov/s/c/remarks/135352.htm.

3. Matthew Lew, "Hillary Clinton, E-Diplomat, Embraces New Media," Associated Press, March 23, 2009, http://www.sfgate.com/cgi-bin/article.cgi?f=/n/a/2009/03/21/national/w064913 D54.DTL&type=tech.

4. Micah Sifry, "Hillary Clinton Launches '21st Century Statecraft' Initiative by State Department," TechPresident.com, May 13, 2009, http://techpresident.com/blog-entry/hillary-clinton-launches-21st-century-statecraft-initiative-state-department.

5. Clinton, "Remarks on Internet Freedom."

6. Vint Cerf, in discussion with the author, February 26, 2009.

7. Clinton, "Remarks on Internet Freedom."

8. "UN Chief Calls for Treaty to Prevent Cyber War," Agence France-Presse, January 30, 2010, http://www.google.com/hostednews/afp/article/ALeqM5h8Uvk-jpSvCWT-bqYSg1Ws 4I4yAA.

INDEX

Abtahi, Mohammad Ali, 192, 193
Advanced Research Projects Agency (ARPA), 29, 30
Africa, future of Internet in, 12–13. *See also* Senegal
Agence de Régulation des Télécommunications et des Postes (ARTP): mismanagement, 92, 101; objectives of, 91; tender for second national operator license, 101
Ahmadinejad, Mahmoud, 195; election of, 195; personal blog of, 195–196
anti-filtering tools: Psiphon, 7, 161; Tor, 6–7, 161
ARPA. *See* Advanced Research Projects Agency
ARPANET, 30, 31
ARTP. *See* Agence de Régulation des Télécommunications et des Postes
Asia, future of Internet in, 12. *See also* South Korea
Asian Financial Crisis of 1997: causes of, 52; impact on South Korea, 52
authoritarian nations, Internet access in, 158, 207

Batik newsletter, 87–88
bloggers, reformist. *See* reformist journalists and bloggers

Cerf, Vint, 205
China: Internet censorship policies, 159; Internet users in, 158–159
Chon Kilnam: academic career, 37, 40–41, 67–68; birth in southwestern Japan, 25; on cable modem service, 49; character traits, 24; on computer networking, *see* computer networking project; development of Korean nationalism in, 27, 28; doctoral students of, 67 (*see also* Hur Jin Ho; Kim Yoon; Lee Dongman; Park Hyunje; Song Jae-kyeong); doctoral study in United States, 29–30, 31; family, 25; first trip to Korea, 28–29; impact of research conducted by, 23–24; meetings with bureaucrats, 37; meeting with Lee Yong-Teh, 33–34; on political turmoil in Japan, 26–27; schooling, 26; technological

ambition of, 24; work at Collins Radio, 30–31; work at NASA's Jet Propulsion Laboratory, 31; work at systems division of KIET, 35
Chosun Ilbo (daily newspaper), 56
Chun Doo-hwan, 35
Chun administration (South Korea): government spending under, 35; on semiconductor production, 37–38
Chung, Chris, 46
citizen journalism, 54–55. *See also* reformist journalists and bloggers
citizenry, and government restrictions, 160
Clean Slate Internet Project, 68–69
Clinton, Hillary: on mobile phone technology, 206; speech on "Internet Freedom," 202, 203
computer development, South Korea: export-oriented strategy for, 35; Lee Yong-The's views on, 33; nationalistic drive for, 35–36
computer gaming addiction: counseling sessions for, 62; deaths caused by, 60–61; diagnosis of, 62; efforts to combat, 61; indicator for, 61; treatment for, 62
computer networking project, 37; goal of, 38–39; Software Development Network, *see* Software Development Network
console game industry, 18
Cuba, Internet access in, 158
cyberattacks: on Estonia, 135–140; impact on pro-opposition websites in Iran, 5; on Iranian government websites, 4–5
cybercafé, *see* Senegal, cybercafé in
cyber-humiliation, 62, 63; issues related to, 63–64; in protests against U.S. beef imports, 64–65
"Cyber Korea 21" program, 50
CyberLouma, 96
cyberwarfare: need for peace treaty to prevent, 207; potential threats of, 141. *See also* cyberattacks
Cyworld, 21–22

Derakhshan, Hossein, 170, 181–182, 192
DFI. *See* Digital Freedom Initiative
dial-up access: cost of, 47–48; problems associated with, 48, 49

ABOUT THE AUTHOR

CYRUS FARIVAR is the host of Deutsche Welle English's internationally syndicated science and technology radio program *Spectrum*. As a freelance technology journalist, he regularly reports for National Public Radio, *The World* (WGBH/ PRI/BBC), and the Canadian Broadcasting Corporation. He has also written for *The Economist, Foreign Policy, Slate,* the *New York Times, Popular Mechanics, Wired,* and many others. He lives in Bonn, Germany, blogs at http://cyrusfarivar.com, and tweets @cfarivar.